The Greening of Agricultural Policy
in Industrial Societies

Food Systems and Agrarian Change

Edited by Frederick H. Buttel, Billie R. DeWalt, and Per Pinstrup-Andersen

A complete list of titles in the series appears at the end of this book.

THE GREENING OF AGRICULTURAL POLICY IN INDUSTRIAL SOCIETIES

Swedish Reforms in Comparative Perspective

David Vail
Knut Per Hasund
Lars Drake

Cornell University Press

ITHACA AND LONDON

First published 1994 by Cornell University Press.

Printed in the United States of America

♾ The paper in this book meets the minimum requirements of the American National Standard for Information Sciences—Permanence of Paper for Printed Library Materials, ANSI Z39.48-1984.

Library of Congress Cataloging-in-Publication Data

Vail, David J. (David Jeremiah)
The greening of agricultural policy in industrial societies : Swedish reforms in comparative perspective / David Vail, Knut Per Hasund, and Lars Drake.
p. cm. — (Food systems and agrarian change)
Includes bibliographical references and index.
ISBN 0-8014-2750-9
1. Agriculture and state—Sweden. I. Hasund, Knut Per. II. Drake, Lars. III. Title. IV. Series.
HD2018.V34 1994
338.1'8485—dc20 93-39131

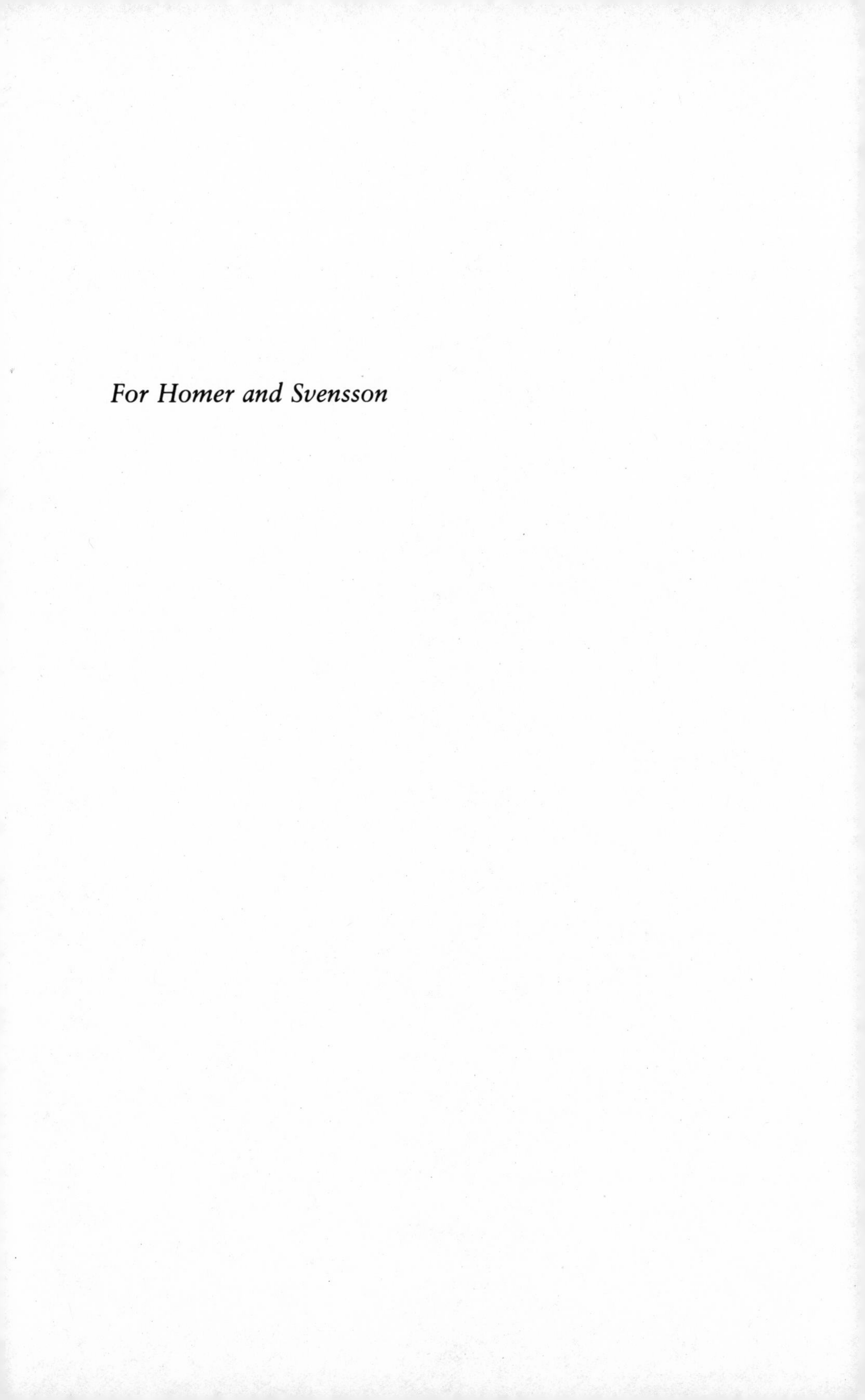

For Homer and Svensson

Contents

Preface

This book grew from two distinct experiences. The first was David Vail's visit to the Swedish University of Agricultural Sciences in 1986. The ostensible purpose of the visit, sponsored by the National Endowment for the Humanities, was to study Swedish policies aimed at preserving and protecting farmland as a food security measure and as a store of amenities and biological diversity. The visit's subtext was an opportunity for a callow American to experience firsthand the rich, varied, and ancient agrarian landscapes of Sweden's Uppland, Dalarna, Södermanland, and Skåne regions. It became vividly clear why "Svensson" (the archetypical Swede) is willing to pay a lot to preserve farmland. The idea of doing a collaborative project on "something about Swedish agriculture" was hatched during a country picnic with Lars Drake at the summer estate of eighteenth-century botanist Carl Linnaeus, a short bicycle ride from Uppsala.

The second experience that prompted the collaboration was our participation in numerous international conferences and study visits and our encounter with the blossoming literature on alternative agriculture, agricultural pollutants, landscape preservation, farm animals' health, and the like. These endeavors heightened our shared sense that something exciting was happening throughout the industrialized world in the 1980s: the greening of discourse about the "agrarian question" and the first piecemeal steps toward an environmentally friendly agricultural policy.

By the end of 1987, after we had met and wandered through farm landscapes on both sides of the Atlantic, we concluded that Sweden was becoming a laboratory for potentially instructive green policy experiments. We decided there was an important comparative story to tell. What convinced us was the Swedish government's decision in 1988 to push ahead with a thorough overhaul of Sweden's archaic agricultural policy regime. The new food policy that emerged in 1990 was hailed as a kind of "magic bullet" that would both exorcise the old regime's costly economic defects and foster environmental protection and provision of agrarian "public goods." If these claims were correct, there was indeed a story to tell scholars, environmental activists, agrarian organizations, and policymakers in all the nations where the state still strongly and visibly regulated the farm sector.

It will be clear to readers, however, that we do not have an idealized or uncritical view of Sweden's new policy regime: we do not present Swedish policy as an ideal model for the world, or even for Sweden! Still, we are convinced that the greening of Swedish agricultural policy can be a source of valuable lessons, both affirmative and cautionary, for other industrial societies.

Many people have served as volunteer consultants on various parts of this volume or as guinea pigs on whom we tested earlier versions. We especially thank Alex Dubgaard in Denmark, Gerhard Haxsen in Germany, David Harvey in England, and Rod MacRae in Canada. To our compatriots on the eastern side of the Atlantic we say, "Tack så hjärtligt!," particularly to Helen Holm, Lena Johansson, Valter Johansson, Lars Jonasson, Michele Micheletti, Ewa Rabinowicz, and Peter Söderbaum; "many thanks" as well to Maureen Hinckle and Willy Lockeretz on the Atlantic's western side.

Fred Buttel, esteemed colleague and editor of the Food Systems and Agrarian Change series at Cornell University Press, made countless valuable suggestions. He also helped keep our spirits up when the manuscript lagged behind fast-moving events because of the press of our teaching, research, and advisory responsibilities. Our collaboration with Fred confirms that economists really can communicate with people from the other social sciences.

On the technical side, we greatly appreciate Irene Hilton's ability to make sense out of an unwieldy bilingual bibliography and Benjamin Vail's skills in designing figures.

We gratefully acknowledge financial assistance and moral support from the Swedish Council for Forestry and Agricultural Research, the Axel Wenner Gren Foundation, the U.S. and Swedish Fulbright commissions, the Swedish University of Agricultural Sciences, and Bowdoin College.

All translations of Scandinavian texts are ours. To the extent that errors of fact and interpretation remain, the finger points directly to us. To the extent that events have overtaken the analysis contained in these pages, our only excuse is *tempus fugit*. The rapid tempo of greening is truly an encouraging sign.

DAVID VAIL
KNUT PER HASUND
LARS DRAKE

Brunswick, Maine
Uppsala, Sweden

Abbreviations

AERI	Agricultural Economics Research Institute (Lantbrukets Utredningsinstitut)
CAP	Common Agricultural Policy
GP	Government Legislative Proposition (Regeringsproposition)
LRF	Swedish Farmers' Federation (Lantbrukarnas Riksförbund)
MCSA	Ministry for Civil Service Affairs (Civildepartmentet)
MOA	Ministry of Agriculture (Jordbruksdepartmentet)
MOE	Ministry of Environment and Energy (Miljö-och Energidepartmentet)
NAMB	National Agricultural Marketing Board (Statens Jordbruksnämnd)
NBA	National Board of Agriculture (Lantbruksstyrelsen)
NEPB	National Environmental Protection Board (Statens Naturvårdsverk)
NM	Nordic Ministerial Council (Nordiska Ministerrådet)
OECD	Organization for Economic Cooperation and Development
SAP	Swedish Social Democratic Party (Socialdemokratiska Arbetarpartiet)
SBA	Swedish Board of Agriculture (Statens Jordbruksverk)
SIFO	Swedish Institute for Opinion Research (Svenska Intsitutet för Opinionsundersökningar)
SLU	Swedish University of Agricultural Sciences (Sveriges Lantbruksuniversitet)
SP	Swedish Parliament (Riksdagen)
SS	Statistics Sweden (Statistiska Centralbyrån)

The Greening of Agricultural Policy in Industrial Societies

1

Agricultural Policy Reform: A Contest of Meanings

By the early 1980s, the handwriting was on the wall: the core agricultural policy instruments used for decades in most advanced capitalist nations had created an economically untenable situation. The triple bind of chronic excess production, escalating public farm expenditures, and depressed international food prices was widely attributed to policy measures that inflated domestic farm prices, restricted imports, subsidized exports, and encouraged ceaseless technical and structural "rationalization." People were also increasingly aware that these policies worsened the environmental impacts of modern agriculture, from monotonous landscapes to polluted water. The heightened tension surrounding agricultural policy in industrial capitalist countries and the articulation of new demands for a resolution of agriculture's environmental contradictions are the setting for this volume.

1.1 Contested Terrain: Political Definitions of Agriculture

In the early years of this century, Karl Kautsky and others used the expression "the agrarian question" to denote the socioeconomic complexities and contradictions of agriculture within industrial capitalism. They well understood that the meaning of agriculture is socially constructed, changeable, and not self-evident. In this book we explore the reframing of the agrarian question in Western industrial nations in the late twentieth century. We investigate the current political context to reshape agriculture's meaning—or rather its multiple and potentially inconsistent meanings—and speculate on how the outcome of the contest will affect the role of the state in agriculture. The stakes are high. The form agriculture takes in the future and its impact on citizens' lives—as consumers, taxpayers, resource stewards, and recipients of environmental amenities and disamenities—depend on what national governments and the international community decide agriculture means to societies that have reached an advanced stage of economic development.

Agriculture's future is debated in newspaper editorials, television features, scholarly conferences, and legislative committees in a score of nations. Internationally, conflict over agriculture's legitimate purposes has been a major stumbling block in trade negotiations. Free trade ideology, and the economic interests it rationalizes, clashes with values as wide ranging as sustaining family farming in economically disadvantaged regions, safeguarding food safety, encouraging environmentally friendly production methods, and protecting the well-being of farm animals. Political actors understand that their ability to influence agrarian symbols is a key to their influence over policy outcomes.

New meanings emerge from the interaction of objective and subjective conditions: a dialectic between immediate events and slowly evolving social structures and human consciousness. From the end of World War II until the early 1970s, the structural and technological transformation of agriculture was more or less in synch with industrial capitalism's overarching economic growth drive and with the prevailing ideology of material progress. Under buoyant macroeconomic conditions, steadily rising agricultural yields, labor productivity, and output were widely viewed as part of the solution to "the economic problem."

Compared with most economic sectors, agriculture received privileged policy treatment during this period. Neoclassical economists (and many others) interpret the extensive state intervention as due fundamentally to the political clout wielded by well-organized farm and agribusiness interests, aided by allies in politics and the state bureaucracy who also have vested interests in continuing intervention (Petit 1985, Winter 1987). The economic motives behind the political agenda of this agroindustrial complex are no great mystery, and the complex's capacity to influence policy is undeniable.[1] Yet to explain

[1] We note, however, that many of the agroindustrial lobby's political victories have been short term or pyrrhic. In particular, policy support has never effectively insulated farmers from the "technology treadmill," the "price-cost squeeze," or the imperative to "get big or get out."

farm policy solely in terms of "rent-seeking" interest group behavior would be to neglect the legitimation of agriculture's special treatment by citizens. Legitimation, as we shall elaborate below, is only partially a function of agriculture's bread and butter contributions to consumer welfare and national food security; it also rests on modern *Homo industrialis*'s "place in the heart" for pastoral landscapes, family farming, and closeknit rural communities. This remains true in many societies despite the fact that industrialized agriculture increasingly diverges from idealized images of close-knit families, neighborly cooperation, traditional seasonal rituals, and virtuous toil in harmony with nature.

Since the early 1970s, the agrarian question has been reasserted and reformulated. One factor, according to Alessandro Bonanno (1991), is precisely the old policy regime's success in answering the narrow economic version of the question. Rationalization and food security objectives were largely fulfilled; food expenditure's share of income declined; farming as an economic system and as a way of life was increasingly subsumed within large-scale agroindustrial structures; and farm people's national importance dwindled, both economically and politically.

The farm policy debates of the past twenty years have not, however, been about writing the next chapter in an agrarian success story. Instead, they have echoed with a refrain of crisis: "food crisis," "farm crisis," "farmland crisis," "overproduction crisis," "world market crisis," "farm policy crisis," and so on. Whether the term is used literally or figuratively, "crisis" evokes an unsustainable situation, a decisive period, a turning point. The evocation of crisis is in part a tactic in the contest over agriculture's meanings, but the term is ambiguous, since it has different meanings for different actors. The causes, impacts, and cures for the "farm policy crisis" would be viewed quite differently by a commercial farm lobbyist, a small part-time farmer, a legislator from an urban constituency, an environmental activist, and a neoclassical economist.

To make a sound interpretation of agriculture's policy crisis and what is at stake in debates about the appropriate response, we must understand how agriculture and the agroindustrial complex operate as subsystems of larger economic, sociocultural, and ecological systems. Likewise, we need to understand the place of a relatively autonomous agricultural policy within the web of macroeconomic, regional, environmental, trade, and other policies. This notion of agriculture's and agricultural policy's embeddedness in larger social systems is developed in the following section and in Chapters 2, 5, and 6.

Long-simmering disputes about why agriculture's fate should matter to an urban, industrial society have come to the boil in several nations, in the European Community, and in the current GATT (General Agreement on Tariffs and Trade) negotiations. In this period of widespread neoliberal reform, in which the social benefits of competitive markets and the social costs of government failure are heavily emphasized, a central issue is whether distinct farm policies should be abolished and agriculture treated like any other economic sector.[2] In effect, this would mean stripping agriculture of the cultural associations and ethical commitments that have long justified extensive regulation and protection from market forces—at a cost to taxpayers and consumers. An implicit goal of the neoliberal project is to minimize the social value attached to agriculture's roles beyond supplying food and fiber at the lowest possible (short-term) monetary cost. Neoliberal voices may have been loudest and most influential in English-speaking nations, particularly during the Reagan-Thatcher period, but they are in the ascendancy nearly everywhere. Indeed, objective conditions justify many of their criticisms of existing farm policy:

- It is a dubious claim that excess food production, dependent on imported farm inputs, contributes to food security
- Food security should be reconceptualized to fit the post–Cold War environment
- Macroeconomic efficiency would be increased by reallocating labor and capital out of agriculture
- Price supports act like a regressive tax on consumers, one that disproportionately benefits large-scale agricultural producers
- With agriculture a small and shrinking part of the total food economy, import protection has increasingly become a covert subsidy to the agribusinesses that surround farmers on the input and output sides
- Highly industrialized farm operations, typified by hog and egg factories, bear little resemblance to the hallowed family farm
- Agroindustry's public relations organs systematically exaggerate the social benefits of farm programs and the catastrophic consequences of changing them.

The neoliberal objective is to shift the boundary between the political and economic spheres, and one of its tactics is to challenge popular images previously used to justify agriculture's privileged treatment. As

[2] The term "neoliberal" is widely used in Europe to describe a stance that is skeptical about the state and optimistic about market forces. In American political discourse, the term "neoconservative" would be used to describe this orientation. See Buttel 1992.

Paul Stern (1990) stresses, resetting boundaries inevitably redistributes power. From the neoliberal perspective, farmers and agribusiness have too much power. They should be subjected to a healthy dose of market competition, particularly international, just as producers of bicycles and banking services are.[3] This doctrine is not, of course, a recent neoliberal invention; it traces back to Ricardo, Mill, and the nineteenth-century economic liberals.

Michel Petit (1985) has shown that general economic conditions have been the prime movers of major agricultural policy reforms in the present century. This was true in both the economic depression of the 1930s and the period of international commodity inflation cum domestic stagflation of the mid-1970s. Both these economic crises led to increased state support for farmers, whereas at the present neoliberal moment, economic and ideological pressures support the reduction, if not elimination, of agriculture's special supports.

The neoliberal critique was decisive in reopening the agricultural policy debate, yet the neoliberal position has neither monopolized discourse nor succeeded in selling a narrowly economistic meaning of agriculture to the electorate. The competing viewpoints are grounded in venerable traditions of agrarian advocacy, such as national food security, resource conservation, and rural socioeconomic revitalization. Greening incorporates elements from all these traditions.

1.2 Greening: New Demands and Expectations

In the 1980s, environmental protection became a mainstream political issue and the environmental movement came of age. Political parties and candidates for election revised their propaganda to convince voters of their environmental commitments, while corporations redesigned their packaging and advertising (and sometimes their products) to project an environmentally friendly image. Industrialized food and agricultural systems, specifically, became a focus of public anxiety and political demands. Environmental organizations and the mass media publicized global developments with ominous potents for agriculture—the greenhouse effect, ozone depletion, desertification—and they riveted attention to agriculture's detrimental effects close to home:

[3] To argue that farmers should be forced to compete like everyone else overlooks the fact that in capitalist economies, many economic sectors have some characteristic—and change-resistant—forms of state support.

pesticide residues in groundwater, topsoil erosion, eutrophication of lakes and streams, destruction of wildlife habitat.

From England to Japan, an important new aspect of agrarian politics in the 1980s was the growing involvement of environmental organizations and their incipient alliances with the alternative agriculture movement and more traditional agrarian and historic preservation groups. Rejecting a narrowly economistic construction of agriculture, these "new social movements," as Fred Buttel (1992) terms them, sought to infuse political discourse with "green" values and meanings that may loosely be categorized as follows:

- Nature conservation[4]
- Preservation of open and varied farm landscapes
- Reduction of environmental pollution
- Protection of human health
- Humane treatment of farm animals
- Sustainable resource use.

Buttel offers a useful gloss on the notion of greening: "By greening, I mean the processes by which environmental concerns are nurtured within social groups and modern environmentally related symbols become increasingly prominent in social discourse. Greening can thus be taken to be a broad social force, equivalent to, for example, the Protestant ethic or the formation of an oppositional working class culture in earlier historical periods" (1992:1–2). At this point, greening remains somewhat amorphous, both as an ideology and as a social movement, and only time will tell if greening will have a pervasive long-term influence comparable to Protestantism and the industrial labor movement.

Buttel's reasoning, explored at length in this book, is that ideological greening is not sui generis, but rather an endogenous response to negative symptoms and their underlying socioeconomic causes. In obvious ways, greening is a reaction to fast-accumulating symptoms of environmental destruction, a few of which have already been mentioned. At a somewhat deeper level, it is a response "to the institutionalized rationalities and the practices of centralized bureaucracies of the state and economy that lead to technological, health, and environmental risks" (Buttel 1992:2). Outside agriculture, these institutionalized rationalities and practices are exemplified by the large-scale production of

[4] The ambiguity of meanings in agrarian discourse is suggested by the use of "nature" and "nature conservation" to describe distinctly human-influenced agroecosystems.

weapons systems, by planned obsolescence as a key element in capital accumulation, by heedless use of earth, air, and water as waste sinks, and by nonsustainable fossil fuel dependency.

At the most fundamental level, Buttel locates the greening of ideology and political practice within the "larger currents of social change in the late twentieth century" (1992:2). Indeed, the present phase of neoliberal political hegemony is itself the result of postwar capitalism's long wave of economic expansion and its subsequent crisis. (This interpretation is elaborated in Chapter 2.) Ideological greening and the new green social movements can be interpreted as a countervailing force to neoliberal ideology, with its narrow notion of rationality, and the vested interests it explicitly or covertly promotes. A central theme of this study is the uneasy coexistence of political conservatism and environmentalism. It should be recognized, however, that many green ideological adherents and green organizations are difficult to pin down on the left-right spectrum that was used to describe political positions in earlier stages of capitalism. Greenness cannot simply be equated with progressive social and economic convictions.

"Sustainable agriculture" has become a buzzword and yet another contested meaning. The term was initially used by the alternative agriculture movement to highlight industrial agriculture's environmental degradation and to affirm the need for resource conservation in a world of growing scarcities. As Richard Norgaard (1991) points out, sustainability is in essence an ethical and ecological concept, connoting the present generation's moral responsibility to future generations and to perpetual renewal of basic biological and physical capacities for life sustenance. But the rhetoric of sustainability has long since been appropriated by conventional farm organizations and agricultural bureaucrats to promote their own agendas. Thus the language of sustainability has been invoked to justify acreage set-asides and other input-reduction schemes whose main purpose is to cut output and raise depressed farm incomes. Such attempts to co-opt a potent metaphor presuppose an environmentally literate public (Buttel 1990).

Throughout advanced capitalism, growing evidence that agriculture's trajectory toward industrialization cannot be sustained has been met by policy initiatives as well as rhetoric. The responses of politicians and economic interest groups typically reflect a mix of genuine convictions, conflicting goals, and tactical opportunism. Their quest seems to be to discover "magic bullet" instruments that curb the old policy's economic contradictions and protect commercial farmers' material interests while also satisfying at least some green demands. A statement by the secretary to the parliamentary group that designed Sweden's 1990 food policy reflects this attitude: "We face a double challenge that must be met. It means being able to limit [economic] support to agriculture and to do it in ways that contribute to a better environment. Regarding the environment, we have the ambition to go a step further than simply limiting negative effects. We want, as well, to contribute to agriculture's positive environmental value" (Sohlman 1989; translated by the authors).

As policy measures and scientific knowledge to promote sustainable agriculture are developed, there is an obvious tension between two quite different conceptions of sustainability. These might crudely be labeled the ecological and the technological—or holistic and reductionist—approaches. They appear to reflect distinct social ideologies as well as divergent philosophies of experimental natural science. Pragmatically, this translates into a split between those in the alternative agriculture movement who advocate a fundamental transformation of industrialized farming systems and those who support incremental refinement of precisely targeted biological, chemical, and mechanical solutions to narrowly defined agroenvironmental problems.

There is some common ground between the two camps, but a deep cleavage is revealed by three controversies that have assumed international proportions: the use of steroids in beef production, the commercial diffusion of bovine growth hormone, and the development of herbicide-resistant crop strains. The economic stakes are high for input supply corporations and for producers of beef, milk, and crops in many nations, but the disputes also reflect philosophic disagreement between the ecological and technological views of agriculture and political conflict over the power that agribusiness corporations and competitive markets have to dictate the pace and direction of technological change.

The absence of either an intellectual or a value consensus about the meaning of sustainability was reflected in an American study, *Alternative Agriculture,* released in 1989 with the imprimatur of the National Academy of Sciences. On the one hand, the report affirmed both the environmental and economic benefits of using complex crop rotations and cultural practices to maintain soil fertility and limit dependence on chemical biocides. This was widely interpreted as a vindication of the alternative agriculture vision. A close reading of the report's case studies of "successful" farms, however, reveals a contradictory "subtext" that validates industrialized agriculture. Many of the farms are highly specialized, extremely large scale, and dependent, at critical junctures, on synthetic fertilizers, chemical biocides, and "finely tuned" biotechnologies to cope with the fertility and pest- and weed-control problems that tend to arise in monocultures (Gianessi 1989, Jackson 1989, National Research Council 1989).

Most public research funding continues to be allocated to narrow technical problem solving rather than agroecology, yet agroecology research is particularly dependent on public funding, since it tends to have relatively little commercially appropriable "output" to attract private investors. Perhaps most important for the future meaning of sustainability, corporate research and development vastly overshadows state-financed research in cutting-edge biotechnology. In aggregate, the resources channeled to narrow technical solutions are several times those allocated to broad ecological approaches (Lacy et al. 1988). Buttel interprets the political economy behind this pattern thus:

> Technical and scientific politics will play a major role in shaping the course of broader sustainability politics. Research policy, which in turn shapes technical possibilities, tends to reflect prevailing patterns of social power—that is, science tends to serve dominant groups. Yet scientific institutions, which are legitimated on grounds of objectivity, occasionally can exhibit the autonomy implied in this ideology. The political pressure placed on agricultural research officials over the past decade, much of it on sustainability grounds, has had some modest effects in this direction. (1990:8–9)
>
> One of the chief barriers to achieving agricultural sustainability will be the private character of the agroinput supply industry . . . input suppliers will be interested in developing new value-added inputs which, in total, involve expansion of farmer dependence on purchased input commodities. . . . The technical side of agricultural sustainability will [thus] depend on whether fiscal and political conditions permit a renaissance of public agricultural research. (1990:14)

From our viewpoint, a green agricultural policy need not deny economic priorities or the constructive role of markets in allocating production resources and correcting policy defects. But greening does attempt to balance the utilitarian ethos and narrow efficiency criteria of mainstream economic thinking with more encompassing values and success indicators. The challenges, it seems to us, are to respecify the broad social meaning of agriculture for the late twentieth and twenty-first centuries and to reshape policy regimes that have proven to be defective by all relevant criteria, whether economic, cultural, ethical, or ecological.

In the latter 1980s, green agricultural policy initiatives sprouted up in every advanced capitalist country. They were frequently reactive, hasty, and piecemeal—not patiently crafted, thoroughly tested, and internally consistent. This book deals principally with the greening process rather than economic efficiency, yet we note with concern that the choice of agro-environmental measures seldom seems to be guided by systematic analysis of the economic consequences. The following chapters make plain three convictions about the place of economics in agroenvironmental policy reform: First, there is no practical alternative to markets as a guide to agricultural resource allocation, even though intervention is needed to correct many forms of market failure such as excessive pollution, undersupply of public goods, and undervaluation of the rights of future generations. Second, in a world of scarce resources, economic tools are very often useful in constructing cost-effective policy instruments, even when their goal is environmental or cultural. Third, in most advanced capitalist nations, the accumulated irrationalities of existing agricultural policy are so pervasive that it is possible to invent a new policy regime that improves environmental and economic performance simultaneously. In section 1.4 we sketch the kind of economic analysis that can best facilitate the process of invention.

1.3 Why Sweden?

As the international dialogue about sustainable agriculture intensified in the 1980s, Sweden, with little fanfare, implemented a broad range of green initiatives. Several of them—taxes on polluting inputs, financial assistance for conversion to organic farming, and measures to improve farm animals' well being—were quite innovative. And as GATT disputes over trade-distorting farm policies heated up, Sweden quietly reformed its archaic, heavily regulated agricultural regime in 1990. It became the first highly protectionist nation to set a firm timetable for deregulation of agricultural markets.[5] The new food policy was also noteworthy from a green perspective, since price deregulation reduces fertilizer and biocide intensity, agricultural bioenergy produc-

[5] Earlier, New Zealand and Australia dismantled their agricultural import restrictions and price supports, but their previous regimes were not nearly so protectionist as Sweden's. In fact, one of their objectives was to capitalize on an international competitive advantage in livestock products. There was no comparable payoff to induce Sweden's new policy (though there was a large potential payoff to consumers).

tion is encouraged, and targeted land compensation payments protect some of the biological diversity and amenity values of agrarian landscapes.

As a minor player on the world food scene, Sweden has not attracted much attention for its agricultural policy. Its commodity surpluses represent only about 1 percent of the total farm exports of the industrial nations. Sweden's dubious distinction is that it invented the use of variable import levies in the 1940s, as a way to protect farmers from external competition and price fluctuations. Perhaps the best-known fact about Swedish farming is that it is among the world's most heavily subsidized, ranking behind only that of Japan, Switzerland, Norway, and Finland in the OECD's index of producer subsidy equivalents.[6] Visitors from the European Community and North America cannot help but note Sweden's astronomical food prices.

The country's reputation for policy innovation and competent public administration is associated not with agriculture but with a half century of social experimentation in such fields as universal child care, health and pension entitlements, active labor market policy, worker-management codetermination, and full-employment macroeconomic policy. Ironically, all these mechanisms for mediating market forces are currently being reformed along neoliberal lines. Indeed, on the eve of the 1990 food policy reform, the GATT Secretariat praised Sweden for its market-oriented economic policies while criticizing its interventionist farm policy (Foreign Affairs Ministry 1990). Ironically, some east European leaders, seeking a middle way between state centralism and laissez-faire capitalism, have taken an interest in the Swedish model at the very moment it is under siege in Sweden (see Chapter 5 and Vail 1993).

Sweden's long reliance on variable import levies and its pervasive internal market regulation certainly do not recommend it as a role model for agricultural policy. Yet we would argue that Swedish governments have coped with protectionism's perverse economic effects fairly cleverly. For example, Sweden managed to shift most of its excess capacity from livestock products to grains in the late 1970s and early 1980s by raising the price of grain relative to milk, introducing milk quotas, and offering pensions to older dairy farmers. Thus for the most part it was able to avoid the heavy storage costs for "butter

[6] OECD is the Paris-based Organization for Economic Cooperation and Development, a twenty-four-member group of the industrial capitalist nations. Producer subsidy equivalent is a measure of transfer payments from taxpayers and consumers to farmers as a fraction of farm revenues; current world market prices are the baseline for this estimation.

mountains" and other unmarketable surpluses that have plagued the EC and the United States. Sweden's main green initiatives in the 1980s were the following:

- Environmental taxes on fertilizers and biocides
- Recertification of all pesticides, and mandatory training and certification of pesticide applicators
- Manure storage and spreading requirements
- Cover crop "cross-compliance" as a condition for participation in acreage set-aside programs
- Financial assistance for conversion to chemical-free farming
- Compensation for maintenance of *hagmark,* a special type of wooded pasture
- Capital grants for (re)establishing landscape elements and wetlands
- More humane livestock confinement and slaughter requirements
- Prohibition of preventive antibiotic dosing and growth regulators in livestock feed.

It is not our intention to promote Swedish agricultural policy, before or after reform, as a model for other nations. The recent policy innovations are not socially, economically, or environmentally optimal, and it is problematic whether even the most successful measures should be imitated piecemeal. In fact, we draw mixed conclusions about Swedish agricultural policy, past, present, and planned. In Chapters 2 and 3 we explore similarities and differences in agricultural policies and structures, comparing Sweden with other countries. The point here is that other nations and the European Community have their own uniquely evolved institutions and circumstances and must find their own paths to agricultural reform.

It should also be acknowledged at the outset that the ideas behind several Swedish green initiatives were borrowed from other countries, as we detail in Chapters 7 and 10. Indeed, since the late 1970s, the industrial nations have been a collective laboratory for green agricultural experiments. Because Sweden experimented earlier and more extensively than most others, we have tried to extract some useful lessons from its trials and errors.[7]

Sweden's green initiatives in the latter 1980s do not imply that the governing Social Democratic party (SAP) was motivated by a coherent "ecosocialist" vision. On the contrary, we explain in Chapter 5 that

[7] Two of the authors, Drake and Hasund, have been participant observers in Swedish agricultural policy. They have helped to design some of the environmental measures featured in the new policy, carried out technical analysis for government agencies and commissions, and advised Sweden's largest environmental organization on agroenvironmental issues.

Figure 1.1. A birch and juniper *hage* (photo by Bengt Hedberg, courtesy of *Naturbild*)

the Social Democrats' agricultural agenda was dominated by economic rather than environmental priorities and that the SAP actually dragged its feet on some environmental issues. The environmental initiatives of the 1980s were part of a politics of shifting coalitions that frequently cut across left-center-right ideological lines. The deepening of Swedish citizens' general environmental awareness and concern was instrumental in bringing green issues to the center of the farm policy debate. We explore the special feelings of "Svensson" (the typical Swede) for family farming and the countryside and the influence of public opinion on agroenvironmental policy in Chapters 5 and 6. The Swedish romance of the countryside, which remains strong even in today's highly industrialized and urbanized society, is suggested by the great playwright-novelist August Strindberg, in his homage to the *hage,* a wooded pasture with origins in the Iron Age: "I know nothing so enthralling, and in the same way so genuinely Swedish, yes, perhaps the only thing genuinely Swedish, as the hage" (Edman 1990) (Figure 1.1).

Some readers will undoubtedly conclude that Sweden's political culture and institutions are so eccentric that its policy innovations are irrelevant to any other context. Admittedly, only a few relatively small European nations have governance regimes that closely resemble Sweden's. But other readers, especially advocates of a greener, more cost-effective agricultural policy in their own nations and communities, may find helpful hints in Sweden's recent trials and errors.

1.4 "No View without a Viewpoint": A Few Words on Method

Swedish economist Gunnar Myrdal (1978) rejected pretentions to a value-neutral social science, claiming that "there is no view without a viewpoint." We agree. Our viewpoints on agriculture, the environment, and public policy reflect a methodological pluralism that should be made clear at the outset.

We draw on insights and make use of techniques from neoclassical economics, primarily to assess the relative efficiencies of policy interventions and to estimate certain social impacts. Methodological advances within this paradigm, used judiciously, can contribute to more effective agroenvironmental policy measures and to a better grasp of the role of interest groups in the policy process. Two examples of the former are methods of estimating the social value of nonmarketed goods and services, such as aquatic ecosystems and open landscapes, and methods of evaluating risks such as those associated with a specific pesticide or food security policy. Two examples of the latter contribution are public choice theory and game theory, which construe politics as a strategic contest over payoffs and emphasize political agents' self-interested behavior and the state's role as creator and protector of economic rents. Despite our selective use of neoclassical techniques, however, this book is best understood as a contribution to the widening dialogue with—and against—neoclassical orthodoxy and its normative implications.

Our understanding of social institutions and relationships leads us to reject neoclassical economics' psychological/ideological postulate of radical individualism: the view that all social activity ultimately derives from and can be reduced to individuals' self-interested, optimizing decisions. Swedish social life reveals the great extent to which individuals' values, needs, moral principles, decisions, and actions are all embedded in a historically evolved sociocultural matrix. Of course, that matrix continues to evolve, along with individuals' behavioral dispositions and actions. This viewpoint does not deny a central role to self-interested individual and group behavior in contemporary political economy.

By the same token, class positions and relationships, the point of entry for orthodox Marxian social analysis, are not the most fruitful organizing device for the study of farm economics, agricultural policy, or ideological greening in the late twentieth century. One basic reason is that present-day farm operators, individually and collectively, occupy multiple class positions as property owners, entrepreneurs, managers, tenants, wage earners, and pensioners. For another thing, class relations within the larger agroindustrial complex are, as the term itself indicates, complex. In many nations, including Sweden, farm cooperatives own a large share of agroindustrial capital and farm operators have common economic interests with wage earners in agroindustry. And the global agribusiness corporation, a complex and powerful institution, cannot be understood in purely class terms. Despite these qualifications, however, class interests and relations are important for analysis of the food economy and the state. In Chapter 5, in particular, we stress the conflicting material interests of Swedish farmers, food industries, and wage earners (as consumers and taxpayers) which shaped the food policy debate.

In company with institutional economists and socioeconomists, we view human beings as relatively autonomous individuals who are also members of overlapping collectivities, and self-interested beings who are also motivated by moral convictions and a sense of solidarity with certain others. Amitai Etzioni (1988) uses the expression "I—We paradigm" to describe a mode of analysis that starts from these premises. Each individual is enmeshed in a web of relationships based on kinship, ethnicity, gender, class, nationality, and so on. These collectivities are not only the sites of human activity, but also prime socializing forces. Relationships within them largely shape individuals' identity, values, moral convictions, and behavioral norms. They help explain, for example, why Swedish farm organizations and agricultural policies resemble those of Austria but are quite different from those of the United States (see Chapter 2). In a word, "local cultures" of agriculture are still distinguishable, however attenuated they may have become under the relentless pressures of international competition, technological transformation, and cultural homogenization.

Viewing individuals as moral beings and members of multiple, overlapping collectivities naturally affects our interpretation of economic and political behavior. In particular, both theory and personal experience convince us that the collective actions of, say, a farmers' cooperative, a political party, or a nature conservation society cannot be derived straightforwardly from the members' material interests.[8] This is not just the result of ubiquitous "principal—agent" problems;[9] rather, what we have in mind is the influence of traditions that perpetuate shared values, beliefs, and meanings: in other words, culture. Culture tends to act as an inertial force, retarding institutional and behavioral adaptations to changing objective conditions. Yet in certain circumstances, culture may facilitate change: "Rapid change can happen because people are mobilized by values as well as interests and because the connection between political choices and values is not self-evident" (Stern 1990:43). Greening is a prime contemporary example of a cultural mobilization for change.

Furthermore, interest organizations (both their principals and agents) are socialized—Swedish analysts sometimes say "tamed"—by their participation in stable, ongoing relationships with other interest organizations. Social contracts, often with the state mediating conflicts, are a prime example. In Albert Hirschman's (1971) terms, a form of loyalty develops, even between organizations whose principals have conflicting material interests.

Two notions are conflated in the preceding discussion. One notion is that individuals' pursuit of self-interest is mediated by culture, loyalties, and moral convictions. Etzioni is on the mark with his criticism of neoclassical economics' tautology that "however seemingly unselfish the goal, people are motivated by their own interests in pursuing it" (1990:31). The other notion is that "human sentiments" are endogenous and therefore mutable and subject to manipulation. Mutability is illustrated by the rise and fall of Swedish citizens' concern about environmental problems, which has occurred several times since the early 1960s. Manipulation is exemplified by the farm lobby's media campaigns to mold a public opinion that is tolerant of economically costly farm programs. In conceptualizing the state and public policy, we explain in Chapters 5 and 6 why the cultural and moral foundations

[8] The organizational constraints discussed here are distinct from the classic problem of aggregating the diverse interests of members, as explicated theoretically by Kenneth Arrow (1974) and concretely, in the case of the Swedish Farmers Federation, by Michele Micheletti (1990).

[9] The agency problem arises whenever principals, such as dairy producers, enter into relationships with agents, such as legislative lobbyists who represent farmers' interests. At one level, the problem is simply that there is no straightforward way to aggregate the principals' interests (i.e., all dairy farmers do not have identical economic interests). But in addition, the interests of the principals and those of the agent are not identical, for example, when a lobbyist loses "reputational capital" by employing certain measures to promote the farmers' interests.

of agricultural policy cannot be discounted or considered simply another "vector of preferences," as is often done in neoclassical analysis.

Another neoclassical postulate that requires critical inspection is rationality. From the neoclassical viewpoint, individuals with subjective ends (ordered preferences) and limited means (resources) dispassionately choose the means that maximize their utility.[10] Our experience suggests that much individual and collective behavior is nonrational, in the sense that subjectivity (values, emotions, ethical norms, etc.) influences the choice of means employed as well as the ends pursued. When means take on intrinsic value, ends and means cease to be neatly separable categories. There is a tradition in economics, historically associated in the English-speaking world with Veblen, Keynes, Myrdal, and Galbraith, that recognizes how often the choice of means is permeated by subjectivity. Reflecting on agricultural policy in our home countries, we recall many situations in which values held by farmers, politicians, bureaucrats, or citizens at large influenced decisions about policy measures, not just policy goals. The use of price supports to promote the farm income goal, instead of transparent and efficient transfer payments, is perhaps the classic example. Farmers want to be compensated for their productive effort and contribution, not stigmatized as welfare recipients;[11] citizens have by and large been sympathetic to this demand, despite its high economic cost.

As economists, we are not immune from frustration when policymakers reject our suggestions for more consistent and efficient instruments to achieve their stated goals. (This is evident in our critique of Sweden's piecemeal green measures and food policy reform in Chapters 7, 8, and 9.) Swedish policy design has long been a contested terrain, pitting a rationalistic orientation against means-subjectivity. Although Gunnar Myrdal rejected value-free social science as an illusion, he was a leader among Sweden's "rationalizing intellectuals." As early as the 1930s, he criticized Swedish food import restrictions as an economically inefficient and thus irrational way to pursue the valid

[10] In multiperiod choice situations characterized by uncertainty, the rationality postulate is that individuals seek to maximize the discounted present value of expected utility, subject to their risk and time preferences. If, in addition, information is limited, distorted, or costly, the rationality postulate is modified further to stress "process rationality." Imperfect information makes it impossible to determine the optimum action (instrument) unambiguously, but the rational actor acquires and processes information efficiently in choosing among instruments (Elster 1989).

[11] This preference is consistent with farmers' tactical motive for advocating price supports: they are less visible and thus politically more secure than explicit income transfers financed directly from the state's budget.

social ends of farm sector stability and distributional equity (Eyerman 1985, Thullberg 1983).

In the 1990s, there is a need for critical rethinking about what constructive role the economic conception of rationality can play in designing sustainable agriculture policies. Norgaard offers the following cautionary note:

> Economists need to explicitly recognize that sustainability is an *equity question* being debated in various moral discourses utilizing *ecological reasoning* and that sustainability will be chosen through *politics*. Economists in this framing can inform the political process of the impacts of different equity decisions and the most cost-effective ways of reaching them. Economics can interact with moral discourse, environmental lines of reasoning and the political process, but cannot "rationalize" them. (1991:5; emphasis added)

Finally, our approach emphasizes both social and environmental evolution: dynamic processes that neoclassical economics does not handle very effectively. Norgaard exaggerates only slightly in claiming, "Neoclassical market economics is an atomistic-mechanistic model which views systems as consisting of parts and relations between the parts which do not change. Evolutionary thinking looks at systems as undergoing changes in their parts and their relations. One cannot have it both ways" (1991:5). In particular, the real world is not populated by immutable decisionmakers—the individual, the firm, the politician, the bureaucrat—who react to changing external conditions in deterministic ways. At many points we stress that core institutions and relationships, as well as the underlying values and preferences of individual actors, coevolve with their social and "natural" environments.[12]

Finally, contemporary policy analysis urgently needs the sense of history that such scholars as Michel Petit (1985) and Alan Winter (1987) provide. Whether the task is to explain the shift from farm policy inertia to reform or to predict the future trajectory of international agricultural trade, we need to grasp where existing policies come from. How did particular social structures and contingent factors bring

[12] Our way of analyzing and evaluating the interaction between social systems and the ecosystem falls under the heading of social ecology rather than deep ecology. Adherents to "nonanthropocentric" philosophic positions, which hold that nonhuman species have "natural rights" and that the needs of humans should not take precedence over the needs of other species, will undoubtedly have quarrels with the way this book analyzes agroenvironmental problems and policies.

forth those policies? What changes in those conditions are necessary and might be sufficient to overcome policy inertia?

Our reading of the past and our tentative projections of the future thus borrow insights from several disciplines and we attempt no grand theoretical synthesis. The best shorthand label for our conceptual framework is "political economy," and our narrative approach relies on what anthropologist Clifford Geertz calls "thick description." The text is loosely organized in sections as follows:

In Chapters 1 and 2 we introduce the emerging crisis of Western agricultural policy. In Chapters 3 and 4 we provide a brief history of agriculture and farm policy in Sweden. We theorize the political economy of agricultural policy and environmentalism in Sweden in Chapters 5 and 6. In Chapters 7, 8, and 9 we discuss the greening and economic reform of Swedish agricultural policy. And finally, in Chapters 10 and 11 we place the greening of agricultural policy in a comparative perspective.

The book has two fundamental goals. The first is to explain the phenomenon of greening and how it has reshaped agricultural policy. The second is to evaluate green policy instruments and suggest how they might be improved. We pursue both goals by exploring the Swedish experience.

2

Agricultural Policy Crisis in the Industrial Nations

The term "crisis," a leitmotiv of agricultural policy discussions in the 1980s, has taken on many shades of meaning. In popular and journalistic usage, "crisis" covers a wide range of symptoms and problems: from the local to the global, from the immediate present to the distant future, and from the certain to the speculative. Most academic social scientists reserve the term to describe a fundamental historic discontinuity: the cumulative effects of a system's internal, or structural, contradictions. In a world of nation-states and global economic and environmental feedbacks, crisis symptoms may appear to a single nation as exogenous shocks.

Working in this tradition, we view crisis as both a condition evolved from the past and a process leading into the future. Crises can be thought of as decisive periods—possible turning points—in the evolution of a social or ecological system.

Growing concern about sustainable agriculture, grounded in perceptions that present production systems are not ecologically, economically, or socially sustainable, is a thread connecting the popular and the analytic notions of crisis. It is widely sensed that industrial nations, capitalist and former communist alike, have come to the end of an agricultural era: that the food production and trade regime that took shape after World War II is on its last legs, and that archaic domestic policies have become so costly, even counterproductive, that reform is imperative.

The agrarian question has been radically reframed in the crisis era. At the 1974 World Food Conference in Rome, the consensus forecast and the source of great anxiety was a Malthusian specter of world food scarcity. But by 1986, heads of state at the Tokyo economic summit were decrying the "global structural surplus" of farm products; similar statements have been made at successive summits (Hathaway 1987). Between those dates, the OECD nations' grain production grew by 2.4 percent per year while annual demand growth was below 1 percent. Grain reserves more than doubled, from 15 percent of annual world consumption to nearly 30 percent, even though millions of hectares were "set aside." International commodity prices plummeted. From 1981 to 1986, the decline was 30 percent for wheat, 37 percent for maize, 30 percent for skimmed milk powder, 58 percent for butter, and 19 percent for beef. The downward price trend was accompanied by greater year-to-year price instability.

Most analysts interpret more recent contractions in world grain production and stocks, from 1988 through the first half of 1990 and again in 1991–92, as exceptions that prove the rule of chronic overproduction in the advanced capitalist nations (Berlan 1990, OECD 1987a, Sharples and Krutzfeld 1990, USDA 1991, Viatte and Langer 1990).

Excess farm production in industrial capitalism coexists with serious deficiencies in food supply and distribution in the former Soviet Union and much of eastern Europe. Massive chronic malnutrition and episodic famines persist in large parts of the Third World (see section 2.5). Meanwhile, the comparatively small group of Asian, African, and Latin American nations that have managed to produce export surpluses of such staples as wheat, oilseeds, sugar, and beef face depressed international prices and eroding real terms of trade that tend to nullify their outward-looking development efforts. In Jean-Pierre Berlan's words, "We are presented with the old paradox of too much food for the markets and too little food for the people" (1990:205).

A widely cited 1987 OECD study summarized the near consensus among agricultural economists that the industrial nations' tendency toward supply-demand imbalance is structural, persistent, and closely tied to productionist national policies. Its prognosis: "Barring major policy changes or repeated climatic failures in the main producing regions, the underlying trends of consumption and production and the accumulation of stocks make it probable that the markets for the major temperate zone agricultural products will be characterized by widespread and lasting surpluses over the next ten years" (OECD 1987a:65). This remains the official OECD outlook (OECD 1991a, b; Viatte and Cahill 1991).

Since the 1940s, agriculture has been exempted from many of the rules governing international trade. When the GATT was created after

World War II, the United States insisted on its right to protect farmers from foreign competition. Ironically, the United States now leads the charge to eliminate agricultural trade distortions by the end of the 1990s. The United States belongs to and is backed by the Cairns Group of fourteen agricultural exporting nations. From its inception in 1986, the GATT's Uruguay Round took excess food production as a given and made reductions in farm import protection, export subsidies, and domestic price supports priority goals. Amid the trade liberalization rhetoric, however, the basic elements of national and European Community (EC) policies remained unchanged. Even as negotiations got under way, the OECD nations' aggregate farm supports continued to rise, peaking in 1987 at US$173 billion (well over 1% of aggregate GDP). As many readers will recall, the Uruguay Round collapsed in a stalemate at the end of 1990, with the EC unprepared to accept Cairns Group demands for deep cuts in import protection and export subsidies, and the United States determined to hold other areas of trade liberalization hostage to agricultural reform.[1] The implications of the EC's long-awaited measures for farm policy reform in 1992 and the present state of play in the GATT negotiations are analyzed in Chapter 9.

Rising domestic farm support costs in the mid-1980s clashed with the ascendant neoliberal ideology, as well as with growing fiscal conservatism and governmental budget constraints. The agricultural budget of the Common Agricultural Policy (CAP) ratcheted upward from 0.10 percent of EC members' combined GDP in 1967 to a peak of 0.73 percent in 1988. In the depths of a farm financial crisis, U.S. federal farm expenditures reached a nearly comparable level of $26 billion/year between 1984 and 1986, when over three-fourths of net farm income came from government payments (Buttel 1989, Peterson and Lyons 1989). In Japan, Sweden, and a few other small European nations, the burden of farm policy is borne primarily by consumers via high food prices. There the gross cost of agricultural supports reached nearly 2 percent of GDP. Further, a large part of the cost to consumers and taxpayers never reached farmers as income but "leaked" instead into higher input expenditures, land rents, and (in some countries) revenues to the food processing and distribution industries (NM 1989).

Critical scrutiny of farm policy intensified in the latter 1980s largely

[1] Parenthetically, the December 1990 collapse followed an unsuccessful effort by Swedish agriculture minister Matts Hellström to mediate a compromise between the EC and the Cairns Group.

because of these conspicuous costs, but also because of poor goal attainment and cost-ineffectiveness. In David Goodman's judgment, policy measures increasingly contradicted their ostensible goals: "Agriculture in advanced industrial countries now exhibits precisely those characteristics [farm policy] was intended to attenuate: market instability, low returns on capital, falling farm incomes, and farm failures" (1991:60).

Since the most costly types of income support rise with farm output, they favor large landowners in the most productive regions and work against economically marginal farms and regions.[2] Even though farm policy has been justified nearly everywhere as a way to sustain family farms, price supports and the excess production they induce tend to accelerate the demise of full-time family-labor farms: "the disappearing middle." The tendency for high commodity prices to be capitalized in inflated land values also contradicts their income support function, while penalizing tenant farmers and making entry more difficult for younger farmers. If, however, support prices were cut, there could be a massive decapitalization, bringing hardship to farmers whose life savings are tied up in their land and to operators with heavy debt loads based on land collateral.

As mentioned, countersymptoms have appeared twice since 1988, when episodes of depressed world grain production caused a fall in reserves and a rebound in international prices. In the United States, net farm income rebounded to a fifteen-year high in 1990, as the number of financially vulnerable farms fell to half the peak level of 1985/86. The Department of Agriculture optimistically declared that the farm sector had weathered the storm of the mid-1980s and that the farm crisis was on its way to being resolved. In fact, after a second low wheat harvest in 1991, the acreage set-aside program was scheduled to be phased out in 1993 to ensure sufficient supplies for domestic needs and for export (USDA 1990b, 1991).

All told, the twenty-four OECD members managed to cut farm support costs by 18 percent from 1987 to 1989, and for once, the CAP's expenditures actually fell below its budget appropriation.[3] Then bumper harvests in Europe, the United States, and the Soviet Union

[2] Some countries put a ceiling on payments per farm to counter undesirable distributional effects, but large-scale operators, assisted by their lawyers and accountants, have typically found legal loopholes to rearrange land titles and minimize payment limitations. This is one form of the "rent protecting" behavior analyzed in Chapter 5.

[3] OECD analysts attribute the CAP's temporary budget surplus more to exchange rate movements and declining international carryover stocks than to any specific policy action (Viatte and Langer 1990).

in 1990 undercut farm prices and brought support costs nearly back to 1987 levels (Viatte and Cahill 1991).

In view of the sharp fluctuations in major crop yields from 1988 to 1992, it is too soon to know if we are experiencing a turnaround in the underlying supply trend or merely the transitory effects of supply-control measures coupled with atypical weather conditions and exchange rate movements. OECD analysts Carmel Cahill and Gérard Viatte (1993) assess the medium-term market outlook largely in terms of policy effects: "The medium-term forecasts indicate some easing in structural surpluses and some increase in international prices. But these forecasts face numerous uncertainties. They are based on specific assumptions which may have considerable impact on markets: that the reform of the Common Agricultural Policy of the EC will lead to some reduction of subsidized exports of wheat and coarse grains from the Community; that supply control policies in the dairy sector throughout the OECD area be continued; and that market access for beef in Japan and other Pacific countries continues to improve" (1993:6). Any tendency toward a tightening of farm commodity markets could be nullified, however, by a continuation of sluggish macroeconomic performance and income stagnation in the OECD countries and by the efforts of East European and Third World nations to expand their staple food exports.

A different way of interpreting the agricultural crisis is to underline the systemic connection between expansion of large-scale, industrialized food production systems and their negative environmental impacts. In this construction, it is argued that the present system is not ecologically sustainable and that sharply rising real production costs will sooner or later replace current glut, particularly when negative externalities and ecological vulnerabilities are taken fully into account. This conception of crisis is considered in sections 2.3 and 2.4.

2.1 Capitalist Restructuring and the Agricultural Policy Crisis

The agricultural policy crisis should be considered one moment in a more pervasive crisis of capitalism on a world scale, a crisis whose main contours were evident by the mid-1970s and which has been the catalyst for national economic policy reforms and international economic restructuring. Agriculture has been partially buffered from industrial capitalism's trends and cycles by special policy treatment and by distinctive geographic, biological, and social traits that inhibit full integration into the corporate industrial sphere. Yet over the long run, neither policy nor unique sectoral features have insulated agriculture from the forces that have transformed modern capitalism. Since World War II, agriculture has been subject to four developments:

- A "technology treadmill," with continuous pressure to adopt new chemical, mechanical, biological, and, most recently, microelectronic techniques. Input supply corporations have been a driving force behind technical change.
- Concentration of capital at the farm level, through land consolidation and investment in structures and equipment.
- Subordination of direct producers within a corporate-dominated agroindustrial complex.[4] Farmers' entrepreneurial independence has been constricted, and farm value added has shrunk to a small fraction of retail food expenditure.
- Internationalization of input and commodity markets, with such global corporations as International Harvester, Pfizer, and Cargill as leading agents.

Our viewpoint is influenced by analysts who interpret the "long wave" of postwar economic expansion as the result of a historically unique social structure of accumulation: a web of capital-labor-state relationships. Sustained economic growth was greatly facilitated by cheap, abundant fossil fuel and by the weakness of national and international policies to internalize the external costs of industrial growth. High economic growth rates, full employment, rapid expansion of world trade, and increasing international capital mobility were hallmarks of this unique period. At the national level, sustained capital accumulation was typically built on a labor-capital accord and extensive state intervention, which created conditions simultaneously favorable to high profits and rising mass living standards.[5] The state's role centered on Keynesian macroeconomic management, elaboration of entitlement programs (the welfare state), and an array of market regulations and supply-side stimuli, especially public investments in education, physical infrastructure, and research.[6] For more than a quarter century, a growing demand for farm products was ensured by the

[4] In many nations, farmers' organizations own a large share of agroindustrial capital. This does not, however, prevent the vertical subordination of individual farm operators.

[5] This mix of class relations and state roles is sometimes referred to as "Fordism."

[6] In some nations, the state's economic steering has included more detailed sector-level industrial policy or indicative planning.

rising disposable income of all socioeconomic strata and by the spread of diets emphasizing meat and manufactured foods.

The international economic order of the Cold War period rested on the pax americana. The 1944 Bretton Woods meetings created three institutions—the International Monetary Fund, the World Bank, and the General Agreement on Tariffs and Trade—that became cornerstones of postwar trade and finance. The gold-dollar exchange standard provided international financial stability and liquidity. In agriculture, the United States stabilized international prices by selectively accumulating grain reserves and idling cropland (Berlan 1990, Buttel 1989, Friedman 1991, Goodman and Redclift 1989).

The onset of economic crisis is often traced to a conjunction of the 1973 oil shock with secular deterioration in the underlying structures of accumulation. Domestically, nations faced greater inflationary pressure, declining productivity growth, and tightening budget constraints. Internationally, the growing U.S. balance-of-payments deficit undermined the dollar-based exchange system.[7] Massive and rapid movements of liquid capital increasingly nullified national efforts to employ discretionary monetary policy. And core industries of the postwar prosperity were challenged by Japan and the newly industrializing countries (NICs). A major factor diminishing the state's capacity to manage domestic economic affairs was the growing scope and autonomy of transnational corporations, which could increasingly counter monetary, tax, and regulatory policies by moving operations and finance capital across national borders (Crotty 1990, Pollin 1990).

In most of the industrial world, the decade following the first oil shock was marked by more frequent recessions, larger state budget deficits, higher average inflation and unemployment, lower productivity and GDP growth, eroded competitiveness of core industries, and resurgent protectionism. Agricultural policy contributed to macroeconomic problems by feeding inflationary pressure and budget deficits. At the same time, food-exporting nations, such as Canada and Australia, found their exchange earnings depressed by others' surplus dumping.

The protracted economic crisis precipitated a legitimation crisis for

[7] Agricultural policy was also affected by the loss of confidence in U.S. economic leadership, particularly when the Nixon administration reacted to the mid-1970s oil shock and grain price inflation by declaring that the United States would use grain exports as a trade weapon. The United States's commitment to international price stabilization was thus abandoned.

interventionist political parties and governments and reinforced the rightward shift to neoliberal ideology and market-oriented policy. Most governments tried to recreate conditions for rapid capital accumulation and promote industrial restructuring and international competitiveness by adopting some elements of the Reagan-Thatcher model, including abandonment of full employment goals, regressive tax reforms, welfare state cuts, and deregulation of finance and industry (Crotty 1990, Jessop et al. 1991).

It is not yet clear whether the state-capital alliances of the 1980s have succeeded in laying the institutional foundation for a new long wave of capitalist economic growth, much less for a new era of widespread prosperity. In 1990–93, the slow but fairly steady economic expansion of the 1980s gave way to stagnation, and most forecasters anticipate several more years of sluggish growth. There is little doubt that historic changes in international economic institutions and the global division of labor will continue. Despite amendments that weaken the Maastricht Treaty for European Union and delays in its ratification, the EC seems to be moving toward closer economic integration, and it is preparing for the addition of several new members, including Sweden (see Chapter 9). The United States has signed the treaty for a North American Free Trade Area, and Japanese-led integration among the Pacific Rim economies grows stronger. Several NICs are on the verge of becoming full-fledged industrial nations. And the former Soviet republics and eastern European nations are in the early stages of a wrenching transition to open, market-oriented economies.

At the same time, global corporations are rewriting the game rules of capitalism: outflanking national economic policies, dividing and subordinating labor, integrating world financial markets, and engineering a massive concentration of capital via mergers, takeovers, and all manner of joint ventures.

Finally, it seems clear that, whatever the exact contours of the "new world order," resource and environmental limits will be of decisive importance. That fact is dramatized by the 1991 Persian Gulf war, which was largely about control of and access to fossil fuel resources, and by the 1992 Earth Summit in Rio de Janeiro, where agreement was reached in principle, though not in practice, that greenhouse gas emissions must be curbed.

This epoch-making crisis and restructuring of capitalism is the stage on which agriculture's policy crisis is being played out. The economic aspects of the policy crisis, in turn, interact with environmental contradictions and green demands to create the necessary conditions for political change. Macroeconomic conditions and policies affect agriculture via several pathways, with variations from country to country: In the first half of the 1980s, a deep recession followed by a comparatively slow recovery caused food demand to stagnate just when output was growing rapidly. High unemployment rates slowed the absorption of redundant farm labor into nonfarm occupations (and strengthened the EC's resistance to agricultural trade liberalization). Anti-inflationary monetary policy, particularly in the United States in the late 1970s and early 1980s, pushed up real interest rates. Coupled with declining real farm prices, this intensified the financial distress of farmers and rural credit institutions. Fiscal pressures reinforced the neoliberal backlash against farm program expenditures, though in practice most nations and the CAP reacted to the farm crisis by increasing expenditures. Debt burdens and economic stagnation in much of the Third World reduced its food imports and spurred such countries as Argentina, Brazil, and Indonesia to promote staple food exports.[8] Finally, exchange rate volatility reinforced the instability of international farm commodity markets.

David Goodman and Michael Redclift interpret the agricultural policy crisis as an interaction of these broad economic forces with the agroindustrial system's own maturing contradictions:

> Food systems in the post-war period have been increasingly internationalized as a result of the closer integration of national markets, common technologies, more uniform patterns of food consumption, and the overarching strategies of international agribusiness. This integration and interdependence of food systems is a direct result of the post-war internationalization of production and accumulation in the world economy. [But] it would be misconceived to see the "international farm crisis" merely as a conjunctural or cyclical aberration, when it is a logical consequence of the restructuring of the global food economy. (1989:3)

They construe chronic overproduction, a crisis symptom, as a result of the international spread of the "American model" of input-intensive, feed- and livestock-based agroindustry.[9] For a few decades, demand growth counteracted the expansionary impetus of commercial

[8] Brazil exemplifies the nations that are promoting agricultural exports at a heavy environmental cost and in spite of widespread malnutrition among their populations.

[9] It is debatable whether this represents a distinctively American model, since by the 1950s its core features were apparent, if not fully developed, in Canada, Oceania, and northern Europe (see Bonnano 1989, Byé 1989, Coulomb and DeLorme 1989, Drake 1989).

specialization, mechanization, and biological/chemical innovation. From the end of World War II to the early 1970s, demand was buoyed by population growth, a doubling of per capita meat consumption, and state stockpiling and export-promotion schemes. In most countries, the input supply and food distribution corporations' growth strategies propelled farm-level technological transformation and specialization. But productionist farm policy also played a vital role. Specifically, most national governments and the EC fairly consistently encouraged supply expansion via publicly funded research and extension, price incentives, and direct or indirect subsidies to farm investment (Berlan 1990, Buttel 1989, Friedman 1991). Goodman and Redclift are close to the mark in concluding that farm policy regimes ultimately fell victim to the success of their rationalization measures: "The international farm crisis of the late 1980s is a crisis founded on success at bringing science and technology to bear on the uncertainties of agriculture, and developing policies to provide a more secure social and political context in which farmers can operate" (1989:1–2).

Two major supply-side changes stood the Malthusian prognosis of the 1970s on its head: growing Third World exports of food grains, livestock feed, meat, horticultural products, and sugar; and the European Community's transformation from the world's largest food importer to a major exporter. Brazil is the most conspicuous Third World case: in the decade after 1973, its world market share in oilseeds grew from 11 percent to 26 percent and in beef and veal from 4 percent to 15 percent (Hathaway 1987: 52, 60). The EC's transformation is documented in Table 2.1.

On the demand side, three trends eventually led to stagnation in the advanced industrial societies: dwindling population growth, declining income elasticity of demand, and consumers' health-related dietary concerns. Food consumption per capita has essentially remained at a

Table 2.1. The European Community as food exporter

Commodity	Year of transition to net exporter	1985–86 share of world exports (%)
Wheat	1974	29
Sugar	1977	9
Butter	1977	21
Beef	1980	24
Coarse grains	1983	21

Sources: Hathaway 1987:43–65, Peterson and Lyons 1989:14.

standstill in industrial nations since 1980. International demand, which had propped up food prices and trade volumes in the 1970s, also deteriorated as the stagnant Soviet Union and indebted Third World slashed imports. Thus, between 1981 and 1985, wheat trade declined from 101 to 85 million metric tons/year; coarse grains trade dropped even farther, from 108 to 83 mmt (Hathaway 1987: 52, OECD 1987b).

Orderly trade, as mentioned, was also shaken by changes in U.S. grain reserve policy. This was free market Reaganomics at work, but enormous federal budget deficits also undercut the capacity to finance surplus stockpiles. Indeed, the 1985 Farm Security Act introduced export subsidies to get rid of surpluses. Thus did the United States contradict its free trade rhetoric and become part of the trade distortion problem.

By the start of the Uruguay Round of GATT, the air was filled with charges and countercharges about export dumping. The EC's prohibition on imports of American beef produced with growth hormones threatened to precipitate a U.S.-EC trade war and offered a preview of "green" trade conflicts to come. Meanwhile, fourteen food exporters, North and South, formed the Cairns Group to press for agricultural trade liberalization. In sum, agricultural trade conflicts in the late 1980s confirmed Buttel's description of the conjuncture as "one of negotiation and struggle over the parameters of a new social structure of accumulation and the role of agriculture in that structure" (Buttel 1989:47).

2.2 Four Variants of Crisis and Response

Crisis symptoms and policy reactions have taken many forms. To suggest the range of responses, in this section we sketch unique and common features among four actors: the United States and the European Community, the two largest players on the international agricultural scene; Sweden, a minor exporter with high import barriers; and Australia, a nonprotectionist food exporter and Cairns Group member. Table 2.2 summarizes some indicators of similarity and difference.

United States

The American farm situation in the 1980s has been described as an "ensemble of many crises," powerfully influenced by macroeconomic

Table 2.2. Levels and sources of farm producer subsidies in the European Community, United States, Sweden, and Australia (percentage)

	EC	USA	Sweden	Australia
Level of producer subsidy equivalent[a]				
1979–81	35	16	48	9
1987	49	41	57	11
1989	41	29	52	10
1991	49	27	63	15
1992	47	28	57	13
Distribution of farm support costs (1984–86)				
State budget	75	67	23	NA
Consumer prices	25	33	77	NA

Sources: Cahill and Viatte 1993; NM 1989; OECD 1988a, b, 1992.
Note: NA = not available.
[a]OECD's measure of revenue transferred to farmers by agricultural policies, as a fraction of farm revenues.

policies and conditions (Buttel 1989). Its origins can be traced back to measures introduced to cope with the collapse of world markets and farm incomes in the Great Depression. Instruments used frequently since the 1930s include support prices for milk and "program crops," government purchase and storage of surpluses, deficiency payments when prices fall below targeted levels, and voluntary acreage set-asides.[10] Disposal of surpluses via domestic feeding programs (Food Stamps) and Foreign aid (Food for Peace) have also been used to offset low demand.

The 1980s farm crisis followed a period when high export demand and international prices required little use of price and income supports. Indeed, in the middle 1970s, policymakers congratulated themselves for having designed the ideal agricultural policy: farm incomes and export earnings set all-time records, and budget costs were minimal. Farmers' heavy investment in land and equipment in the 1970s was fueled by an unprecedented mix of high farm cash flow, low real interest rates, and government exhortations to plant "fencerow to fencerow," based on erroneous forecasts of continued food scarcity and high prices. Anticipating a golden age, producers borrowed more

[10] Price supports operate through nonrecourse loans to program crop producers. Borrowers may sell the crop on the open market to repay the loan or alternatively repay in kind at a predetermined loan rate, if it is above the market price. Deficiency payments are made for program crops whose prices fall below target levels set annually by the Congress.

heavily than ever before to finance investment (Hathaway 1987, Riemenschneider and Young 1989).

Contrary to these expectations, real farm income fell precipitously from $70 billion in 1973 to $33 billion in 1976 and $20 billion in 1980. Output surged following the investment boom, with grain production growing at 5.0 percent a year between 1973 and 1982, compared to a long-term trend of 2.4 percent. As previously explained, demand deteriorated, with export commodities especially hard hit. From 1981/82 to 1985/86, wheat export revenues dropped 48 percent and the U.S. market share fell from 48 percent to 29 percent. Similar stories could be told for feed grains, soybeans, and rice. Total farm export earnings plummeted from $47 billion in 1980 to a low of $23 billion in 1986, while carryover stocks doubled from 80 to 160 million tons (Berlan 1990:208, Blandford et al. 1988:135).

Depressed prices, falling land values, and punitive interest rates hit heavily leveraged program crop producers hard. Debt service charges, debt/asset ratios, and farm foreclosure rates all reached unprecedented levels.[11] Export-dependent farmers, and hundreds of rural banks and businesses whose fortunes were tied to them, made the painful discovery that they were hostages to a deadly mix of loose fiscal and tight monetary policy that drove up interest rates, cut exports by overvaluing the dollar, and undermined domestic demand by bringing on a severe recession.

Net farm income fluctuated around depression levels ($15 to $30 billion), with government payments accounting for nearly all net income in some years. The budget cost of farm supports grew sixfold from 1981 to 1986, reaching $26 billion. The rate of increase in producer subsidies was far greater in the United States than in the EC, Sweden, or Australia during this period (Table 2.2) Blandford et al. 1988, Hathaway 1987, OECD 1987a, Reimenschneider and Young 1989).

The Reagan administration's farm bills in 1981 and 1985 rejected the previous administration's fear of shortages as well as its agrarian populism, calling instead for wholesale dismantling of federal farm programs. Given the influence of agroindustrial interests over both parties in the Congress, however, the White House proposal had little chance of being adopted. In any case, the Reagan administration's actions contradicted its rhetoric: by 1983 it had already intervened to

[11] In 1985, debt service reached 17 percent of total farm costs. From 1981 to 1987, an $80 billion decapitalization of land values pushed the sector's debt/asset ratio to 40 percent (USDA 1990a).

restrict output, offering program crop producers "PIK" payments to cut acreage ("payment in kind" from government stockpiles). In the dairy sector, the administration did reduce support prices on federal milk marketing orders, but it also launched a whole herd buyout scheme that got government into the beef marketing business. President Reagan, the free trade ideologue, even authorized import protection for U.S. kiwi fruit producers.

The 1985 Farm Security Act, enacted in the depths of the farm crisis, amended existing loan, target-price, and deficiency-payment mechanisms to discourage production, but it did not eliminate any of the old policy's core instruments. As previously mentioned, the 1985 act also sanctioned export subsidies, which reached nearly $1 billion in 1987.[12] In the interval between the 1985 and 1990 farm bills, the farm economy, farm exports, and federal farm expenditures all improved substantially. But policy modifications were less important than depreciation of the dollar after 1986 and tighter international supply-demand conditions in 1988–90 (Hathaway 1987, Koester 1991, Sharples and Krutzfeldt 1990, USDA 1991, Viatte and Langer 1990).

The 1990 Farm Bill was cobbled together during intense bipartisan negotiations to reach an overall budget compromise. The resulting cuts in program support levels do more to reduce excess production than anything in the 1985 act. Authorized spending on export subsidies and surplus stockpiling was cut deeply, total acres eligible for income deficiency payments were reduced by 15 percent, and the loan rate on eligible crops was set at only 85 percent of the average market price of the preceding five years. Although the 1990 legislation contained no fundamental break with the old policy, its planned cuts in price and income support are substantial and indeed greater than those demanded by the Cairns Group in the GATT negotiations. In the event, the mix of incremental policy changes and poor grain harvests in 1988 and again in 1991 reduced grain and soy stockpiles to only 20% of their 1987 peak. The combination of policy and higher market prices reduced federal farm program outlays by 60 percent. (Bradsher 1991, NCFAP 1991).

European Community

The EC's Common Agricultural Policy (CAP), inaugurated in the 1960s, relies on variable import levies to insulate farmers from low

[12] Our earlier contention that the United States ceased to be the world's residual grain supplier must be qualified, since the U.S. government accounted for 80 percent of the increase

or fluctuating international prices. So long as the EC was a net food importer, the policy's cost was borne primarily by consumers through higher prices. Import duties covered most of the cost of storing and exporting periodic surpluses. The situation was transformed in the latter 1970s as a result of cumulative productivity growth and a policy-induced surge in farm investment (Table 2.1). In the first half of the 1970s, the EC had adopted a productionist response to its perceived vulnerability as a net food importer. Encouraged by EC investment subsidies and predictions of continuing high international prices, farmers more than doubled their capital stock in the 1970s (investment growth averaged 12% per year versus 5% in the 1960s) (OECD 1987a:15).

The EC shifted from being a net food importer to an exporter on an expanding scale. By the late 1980s, aggregate production was 15 to 20 percent above consumption and the EC's share of world food exports had increased to 36 percent (from 24% in 1970).[13] The share of the entire Cairns Group, including the United States, at 38 percent, was only slightly larger. As the EC's surpluses and subsidized exports rose, world prices tumbled, so that expenditure for storage and export subsidies increased fourfold. The EC's total producer subsidy equivalent reached US$73 billion in 1987, well over 1 percent of the Community's combined GDP. The EC was awash in its "wine lake" and buried under its "butter mountain." Moreover, its heavily subsidized sales to the Soviet Union, China, and middle-income developing nations took markets away from less protectionist exporters and provoked their antagonism (Hathaway 1987:76–77, Junz and Boonekamp 1991:12, Ostry 1991:17, Viatte and Langer 1990:5).

Wesley Peterson and Clair Lyons observe that "the EC [seemed] constantly on the verge of running out of money to finance the CAP" (1989:11). Responding to both budget pressure and foreign agitation, the Council of Ministers implemented a series of piecemeal measures to restrict supply and modify the CAP's financing. They included:

- Mandatory dairy production quotas, introduced in 1984
- Reductions in the real grain price in 1987 and 1988, with authority to make deeper cuts if production exceeded a threshold level
- Voluntary (compensated) acreage set-asides, since 1988

in global grain stocks during the international market "bust" of 1984–86 and for 75 percent of the subsequent reduction in reserves when prices rebounded in 1988–89. These were classic price-stabilization measures (Sharples and Krutzfeldt 1990:14).

[13] These export figures include intra-EC trade.

- EC financing of national measures to reduce farm input intensity
- A shift of some budget costs to national treasuries in 1988, to stiffen members' resistance to the CAP's high cost.

As early as 1985, a European Commission Green Paper proposed a market-oriented agricultural policy that would decouple commodity prices from support for farm income and regional development objectives. But despite repeated economic admonitions and political resolutions in favor of market-oriented reform, a policy breakthrough was obstructed by multiple counterforces until 1992. In the Community's consensus and compromise politics, well-organized lobbies representing agribusiness and the eleven million people employed in agriculture (7.5% of the EC workforce) were able to maintain effective veto power against fundamental reform. The more reform-minded nations, especially Germany and Britain, could not overcome resistance from the French and their southern European allies. In addition, high levels of cyclical and structural unemployment throughout the EC, particularly in southern Europe, raised the specter of increased redundancy if reform precipitated a sharp contraction of the farm sector. Furthermore, the CAP's budgetary expenditure declined somewhat in 1988–89, because of supply restrictions and higher international prices, so fiscal pressure for reform was dampened. The EC seemed unwilling to deregulate agriculture apart from a wideranging GATT resolution. Finally, as French president François Mitterand recently affirmed, neoliberal ideology has not triumphed completely: the Common Agricultural Policy rests on shared values that transcend economics, including a commitment to "a certain kind of rural civilization" wherein family farms and economically disadvantaged regions play central roles (Goodman 1991, Nello 1989, Ostry 1991, Peterson and Lyons 1989).

Since 1990 the CAP reform debate has again intensified, in the face of renewed budget cost escalation, a threatened collapse of the Uruguay Round, and more effective organization by promarket proponents at the ministerial level and the European Commission. In mid-1991, the commission, led by Agriculture Commissioner Ray MacSharry, proposed sweeping changes that would cut support prices deeply for most commodities, slash export subsidies, and decouple payments to farmers from output levels. After seventeen months of intense lobbying and debate, the Community's agricultural ministers agreed in May 1992 to a compromise plan that could win the weighted majority required for passage. Their plan would gradually reduce grain and beef prices, limit state purchases of surplus beef, slightly lower milk quotas and the butter intervention price, and set the stage for further price reductions. The package, which includes mandatory acreage set-asides for larger grain producers and acreage-based compensation, is far too complex to be detailed here (see Arnold and Villain 1990, SBA 1992b). The current state of the CAP reform, with its implications for the GATT negotiations and for Sweden and other prospective EC members, is assessed in greater detail in section 9.2.

Sweden

Like the EC, Sweden long employed a system of domestic price supports backed by variable import levies. This mechanism, pioneered by Sweden in the 1940s, was the cornerstone of agricultural policy until 1990. For most of the postwar period, farm prices were based on an income parity formula that passed production cost increases through to consumer prices. As in the EC and United States, farm policy in the later 1970s was strongly growth-oriented. Producer prices were raised sharply to meet the farm income goal and encourage a higher degree of food self-sufficiency. Nominal import protection for farm products reached 60 percent in 1980, twice Denmark's level and equal to that of Italy, the EC's most heavily protected member at that time. The resulting price inflation led government to appease consumers by introducing retail price subsidies, but this tactic merely shifted costs to the state budget, at a time when the deficit was nearing 10 percent of GDP.

As in the EC, high support prices coupled with continuous technological advance brought chronic overproduction by 1980. Sweden, however, used a mix of policy inducements (outlined in Chapter 1) to convert most of its excess capacity from livestock products to grain. This temporarily reduced the expense of surplus storage and disposal. But costs escalated again in the mid-1980s, as steady yield increases brought back milk and meat surpluses and pushed the grain surplus from a few hundred thousand tons per year to more than one million tons. Rising surpluses and plummeting world prices caused a sixfold leap in export subsidies in the first half of the 1980s, from US$50 million to over $300 million per year (Rabinowicz 1991a:19).

Facing an untenable budget deficit, the Social Democratic government shifted more of the policy's burden back to consumers with a stepwise elimination of retail food subsidies; by 1986 Sweden's retail food price index was 56 percent higher than it would have been under free trade. In that year, the total cost of agricultural measures reached

$2.32 billion, or 1.8 percent of GDP. Only Japan, Finland, Norway, and Switzerland had more heavily subsidized farm sectors (NM 1989, Viatte and Langer 1990). Farm price supports, import protection, export subsidies, and monopoly power in the food distribution system came under intense criticism on efficiency, equity, and environmental grounds. Since terminating supports and regulations quickly was not politically feasible, a series of ad hoc measures was undertaken to discourage production and shift more of the export finance burden to producers. These measures created a severe cost-price squeeze for grain producers.

Several of Sweden's ad hoc measures echoed U.S. and EC actions, while others were unique:

- A voluntary acreage set-aside program
- Fertilizer and agrichemical levies to finance grain exports (producers had to bear 60% of the export cost)
- Increased slaughter fees to finance meat exports
- Early retirement incentives for dairy and beef producers
- Voluntary milk quotas (through a two-tier price mechanism)
- Price premia for slaughtering calves and piglets, to induce herd reduction
- Financial support for conversion to chemical-free farming.

By tacking these additional measures onto the old policy regime, Sweden restricted production growth. But administrative complexity was further increased and most irrationalities remained.

Australia

After a protracted political struggle, Australia dismantled most of its farm sector supports and price distortions in the late 1970s. The effect is clearly seen in the producer subsidy figures in Table 2.2. The new policy banked on Australia's international competitive advantage in extensive wool, meat, and wheat production. Since Australia has little international market power, the effects of this export-promotion strategy were heavily dependent on world market conditions and fair play by other nations.

Australia's subordinate position in the global food regime was amply revealed by the roller coaster ride its exports took in the 1980s. Its wheat market share and farm export earnings increased in the first half of the 1980s, while the U.S. dollar was overvalued and the United States embargoed exports to the Soviet Union. Then the dollar fell, tight credit conditions spread internationally, the United States began subsidizing farm exports, and EC surpluses soared. Australian wheat farmers were caught in a severe financial squeeze. Meat producers, who formerly had a solid market base in the Middle East, found demand shrinking as oil revenues declined and the EC bid away customers with its subsidized meat exports. These external shocks reinforced preexisting trends toward greater dependence on debt financing and concentration of farm capital to precipitate a rural social and economic crisis comparable to the one that devastated the American grainbelt.

The governing Labor party had no effective means of international retaliation and chose not to bail out bankrupt farmers. In fact, it moved in the opposite direction, abolishing administered wheat prices and the last supplementary market support for dairy products in 1989/90. In Geoffrey Lawrence's estimation, Australia's agricultural future will continue to hinge largely on U.S. and EC policies. Despite efforts of the Cairns Group (founded in Cairns, Australia) to reduce import protection and eliminate export subsidies, Lawrence foresees "little scope . . . to escape from the cycle of overproduction" that is undermining Australian rural economic and social life (Lawrence 1989:259; see also Viatte and Langer 1990).

2.3 The Greening of the Agricultural Crisis

Thus far in this chapter we have interpreted the agricultural policy crisis in terms of unfolding economic contradictions within agriculture and industrial capitalism more broadly. As we emphasized in Chapter 1, however, the crisis has also taken on multiple environmental meanings, which are reshaping the policy agenda and challenging the economic interest group alignments that dominated farm politics for a half century. If indeed agricultural policy regimes have reached a turning point, environmental contradictions are a catalyst. In short, the economic and environmental dimensions of the agrarian question have become intertwined in the advanced industrial nations. Thus, agriculture has become a focus of environmental politics, and environmental protection has become a focus of agricultural politics. The greening of the farm crisis involves a merging of three tendencies:

- Pollution and resource depletion (especially fossil fuel dependency, topsoil mining, and degradation of aquatic ecosystems), which seem to be inherent in price-driven, industrialized farming systems.

- The long-term consequences of worldwide industrial growth, especially nonagricultural sources of acid rain, ozone depletion, and climate modification, which threaten agriculture's long-term sustainability.
- A deepening public concern about environmental problems of all kinds. This subjective force has transformed objective environmental problems into live political issues.

Neither agriculture's adverse environmental impacts nor their politicization is a new phenomenon. After all, the agricultural crisis in ancient Rome's African provinces stemmed from soil salinity caused by overirrigation. In the present century, Australia, China, the former Soviet Union, and the United States all experienced dustbowl conditions after exploiting fragile soils in semiarid regions. In the 1960s and early 1970s, most industrial nations reacted to mounting evidence of pesticides' destructiveness by banning mercuric fungicides and DDT. And in the past decade, a proliferation of agroenvironmental symptoms converged with citizens' growing awareness and anxiety about the environment to trigger greening.

There is developing recognition that modern agriculture's resource depletion, pollution, obliteration of traditional landscapes, and inhumane treatment of farm animals are all caused in some way by policy: artificially high farm prices that strengthen incentives to specialize and use chemical inputs; structural and technological rationalization measures; publicly funded research biased toward higher yields and chemical fixes for pest, fertility, and livestock management problems; and an unwillingness to make polluters pay for the negative spillovers they cause. Growing popular resistance to the continued depletion of public goods, such as cultural-historic landscapes and wildlife habitat, indicates that new green values have become intertwined with longstanding social and cultural concerns about the fate of rural areas.

It is impossible to avoid the conclusion that some green policy initiatives have been adopted in recent years as economically cheap and politically low-risk ways to cut excess farm production. The EC pays member nations to pay their farmers to reduce input intensity, Quebec subsidizes conversion to chemical-free farming, and the U.S. Conservation Reserve Program "retires" erosion-prone land from program crop production. Buttel terms these policies "supply control sustainability" (1990:9).

It is problematic whether a happy marriage can be consummated between agricultural policy's economic and environmental goals. In the United States, for example, enormous federal budget deficits and two years of declining grain reserves provided the justification for drastic funding cuts for sustainable agriculture programs in the 1990 Farm Bill. In the European context, Alessandro Bonanno predicts:

> Future European agriculture will depend largely on the solutions adopted to address environmental problems. The increased political power of environmental groups testifies to the centrality of this issue in the European political arena. Green parties have gained significant electoral power in recent national as well as EEC parliamentary elections. Furthermore, other political formations have included the protection of the environment as a central item in their political platforms. *These changes have strengthened an environmentally oriented agenda which clashes in several instances with old productivistic schemes implicit in current agricultural policies.* The confrontation between these two different orientations and the social and political groups to which they refer will undoubtedly shape the next Europe. (Bonanno 1989:7; emphasis added)

We would stress the intense "clashes" between narrow vested interests and general social welfare, between instrumental rationalism and other modes of knowing and valuing, between fragmented and holistic world views, and between short-term and multigenerational time perspectives.

The 1992 Earth Summit in Rio made it clear that henceforward all aspects of international relations must take environmental goals and problems into account. Indeed, the Punta del Este Declaration, which launched the current GATT negotiations, though best known for emphasizing elimination of agricultural trade distortions, also acknowledged that some forms of domestic market regulation and import restriction should be tolerated as valid ways to protect the environment, animal health, and food safety. In Chapters 9 and 10 we consider why harmonizing such green measures is likely to be one of the most difficult trade challenges of the 1990s.

2.4 Malthusians and Prometheans: Policy Premises for an Uncertain Future

As we have seen, most economic forecasters expect excess production, and its accompanying effects of low farm profitability and trade distorting policies, to be the agricultural syndrome of advanced industrial nations for some years to come, although a GATT agreement and implementation of the EC's recent reforms would reduce these tendencies. The expectation that supply growth will continue to outstrip demand is reinforced by the current efforts of several eastern European and Third World nations, hard pressed for convertible currency, to expand food exports. On the demand side, it is also unlikely that the former Soviet republics and food-deficient Third World nations can absorb enough surplus to offset stagnant or declining international food prices.

Some see a different picture in the recent evidence. Lester Brown and the Worldwatch Institute, for instance, stress that per capita grain production peaked in western Europe in 1984 and in North America back in 1981. Worldwide, they note, grain consumption exceeded production for three successive years between 1988 and 1990, and 1991 marked the largest global decline in grain production per person—6.4 percent—ever recorded (Brown 1991, 1992).

As Petit's (1985) historical studies convincingly show, agrarian politics is normally dominated by short-term forces: immediate economic conditions, interest group demands, and election cycles. On the presumption that this will remain true, we expect the overproduction syndrome to continue to shape the near-term political agenda in industrial nations.

The problematic of sustainable agriculture is not short-term and national, however, but multigenerational and global. The emergence of sustainable agriculture as a metaphor and a rallying cry reflects growing skepticism about the long-term feasibility of industrialized food production in the affluent North and growing criticism of its spread to Asia, Africa, and Latin America. We would like to reflect briefly on the "glut versus scarcity" aspect of sustainability.

The neo-Malthusian premise is that food-scarcity already exists for nearly one billion people and a food-scarce future is in store for millions or billions more because of several convergent forces: continued high levels of human fertility and built-in population momentum; depletion of agricultural resources and the loss of land and water to competing uses; a shrinking backlog of new agricultural technologies; and negative feedbacks from both agricultural and nonagricultural pollutants. These broad tendencies interact with distributional inequities to cause severe chronic malnutrition and with episodic natural calamities to produce mass starvation. (For the forseeable future, the hunger problem in the affluent North is a distribution problem, not a production problem.) The Worldwatch Institute interprets some recent evidence in stark terms:

> Nearly all forms of global environmental degradation are adversely affecting food production: Soil erosion is slowly undermining the productivity of an estimated one-third of the world's cropland. Deforestation is leading to increased rainfall runoff and crop-destroying floods. Damage to crops from air pollution and acid rain can be seen in industrial and developing countries alike. Data from experimental plots indicate that yields of some crops, such as soybeans, are reduced by the increased ultraviolet radiation associated with stratospheric ozone depletion. Waterlogging and salinity are lowering the productivity of a fourth of the world's irrigated cropland. Global climate change in the form of hotter summers may also be taking a toll. (Brown and Young 1990:59–60)

Brown and Young stress that these effects are not and will not be limited to poor nations. Brown contends that U.S. crop yields have already been reduced 5 to 10 percent by air pollution. And William Cline estimates that in the United States alone, heat stress and drought caused by a doubling of atmospheric CO_2 from preindustrial levels will have a negative impact of $18 billion per year on agriculture by the middle of the coming century (Brown 1991, Cline 1992).[14]

A mix of supply constraints and population-driven demand pressures is expected to raise the real resource cost of food, detracting from material living standards everywhere. Among economically marginalized Third World populations, the Malthusian checks of war, famine, and pestilence could strike with a vengeance.

The contrasting Promethean premise holds that technological ingenuity and institutional adaptation can, and in most of the world will, sustain adequate growth in food production to feed a larger world population and do it without any great increase in real resource costs or environmental degradation. Technological innovations, induced by market signals and appropriate policy measures, can conserve resources and internalize many of agriculture's negative environmental effects. Indeed, in the best-case scenario, agriculture will help solve broader resource and environmental problems: as a carbon sink via agroforestry, as a site for recycling urban wastes as soil amendments, and as a source of renewable biofuel and chemical feedstocks. The

[14] Cline's central estimate of $18 billion/year, based on 1990 U.S. production levels and costs, represents nearly 10 percent of U.S. gross farm output. Using simulations from several global climate models (gcm's), Cline forecasts that if CO_2 doubling is reached by the year 2025, the full economic impact would be felt by roughly 2050. The gcm's predict, however, that mean surface temperature will continue to rise for decades or even centuries beyond 2050, intensifying adverse impacts on agricultural production systems (Cline 1992).

technological optimists, as Trgyve Haavelmo and Stein Hansen (1991) point out, tend to emphasize production capacity rather than distribution, making no claim that adequate global output will guarantee adequate nutrition for all.

Price signals and economic incentives operating through competitive markets are crucial to this scenario, and Prometheans tend to be strong advocates of capitalism as a superior form of economic organization. (Until recently, Prometheans were prominent in socialist nations as well, but at the moment, there are no obvious alternatives to capitalism for industrial societies.) They conclude that the greatest untapped agricultural potential is in the Third World and eastern Europe, since state socialist, peasant, and semifeudal agricultural systems have massive inherent inefficiencies that can be corrected with conversion to capitalism.

Promethean scenarios are not typically laissez-faire, however, and it widely is accepted that the state's "visible hand" must help to shape a sustainable agriculture. Governments must foster appropriate scientific and technological development, internalize environmental externalities, and intervene to correct food distribution inequities. There is a common conviction that in capitalist democracies, appropriate private and state responses to changing economic and environmental conditions happen in time, in the right direction, and in sufficient magnitude. Thus Pierre Crosson, considering the implications of global climate change, is optimistic about "the time available to make adjustments and the demonstrated ability of mankind to develop agricultural systems that are productive under climatic conditions even more variable than those portended by a doubling of atmospheric CO_2" (Crosson 1986:110). There is no sense here of any risk that agroecosystems could be stressed beyond some (unpredictable) threshold, precipitating environmental and social chaos.[15] (See Crosson 1986, 1992, Johnson 1984, Nordhaus 1991, and Revelle 1984 for Promethean perspectives.)

Our interpretation of the evidence suggests that the neo-Malthusian prognosis could prove to be tragically accurate for such underdeveloped agrarian nations as Malawi and Bangladesh. We accept an argument made forcefully in the Bruntland Commission report, *Our Common Future,* that for both ethical and self-interested reasons, the advanced capitalist nations must assist Third World nations to make the transition to a sustainable agricultural development path (WCED

[15] An even stronger optimism is held by the prominent American economist William Nordhaus: "Climate change is likely to produce a combination of gains and losses with no strong presumption of substantial net economic damages" (1991:933).

1987). If we focus on agriculture within the advanced capitalist nations, however, it seems to us that the neo-Malthusian prognosis is excessively pessimistic regarding technological and institutional adaptability to environmental and resource constraints. The Prometheans correctly stress the capitalist mixed economies' revealed capacity for crisis management and institutional adaptation. There are many indications, ranging from scientific developments in biotechnology and agroecology to policy measures promoting reduced input and organic agriculture, that piecemeal adaptations are already under way and their tempo is accelerating. Buttel wryly observes that such incremental reactions by agroindustry and the state make "'unsustainable agriculture' . . . potentially 'sustainable' for quite some time" (1990:5).

Although Prometheans can point to evidence of the capitalist democracies' historic success in solving the problem of material scarcity, we question the ethical and value premises that are often implicit in the Promethean position. One is an economistic conception of sustainability that gives little weight to such values as the intrinsic worth (or "existence value") of biological diversity, the preference of millions of people for agrarian livelihoods, and the cultural-historic and amenity values associated with farming and farm communities. A second is the use of a high discount rate to evaluate welfare effects on future generations. And a third is the tendency to downplay risks stemming from long-term, macroenvironmental changes, especially ozone depletion and climate change. Crosson's expectation that the earth's climatic future will be a smooth and gradual extension from the present could prove wrong. Delayed action to limit greenhouse gas emissions, based on a presumption of ample lead time and adaptive capacity, could impose an enormous burden on future generations (see Ruttan 1990).[16] Our bias, in sum, is to take the value of long-term sustainability seriously as a motivation for today's agricultural policy.

2.5 From Crisis to Reform?

Petit's historical studies show that major agricultural policy reforms have been brought about by a conjunction of intractable agricultural problems and powerful pressures originating in the larger political economy. By these criteria, the 1980s seemed ripe for agricultural

[16] In his more recent writings on climate change, Crosson has shown a greater awareness of and concern about the risks, potential costs, and long-term effects of global warming (see Crosson 1989).

policy reform. As Buttel noted in 1989, "never since the Great Depression has there been such widespread attention to farm problems among all the advanced countries of the world" (1989:47). The OECD echoed this theme: "In view of the close links between agriculture and the rest of the economy, there are increasing pressures from other sectors *(especially those concerned with the environment and the alternative use of rural resources)* to broaden the scope of the debate on agricultural policy" (1988b: 18; emphasis added).

The convergent pressures for reform sketched in this chapter are both domestic and international, economic and environmental, social and political. Figure 2.1 shows five broad types of pressure impinging on agricultural policy. What the figure cannot show is the complex interrelationships among all these forces—for example, how environmental degradation and uncertain sustainability of global food production have influenced citizens' green consciousness and provoked their demand for food policy changes.

There was a great deal of reactive tinkering with elements of farm policy in the latter 1980s. But despite repeated ritualistic affirmations of the need for market-oriented farm policies and for international trade liberalization—by the OECD trade ministers, the G-7 heads of

Figure 2.1. Convergent pressures for agricultural policy reform

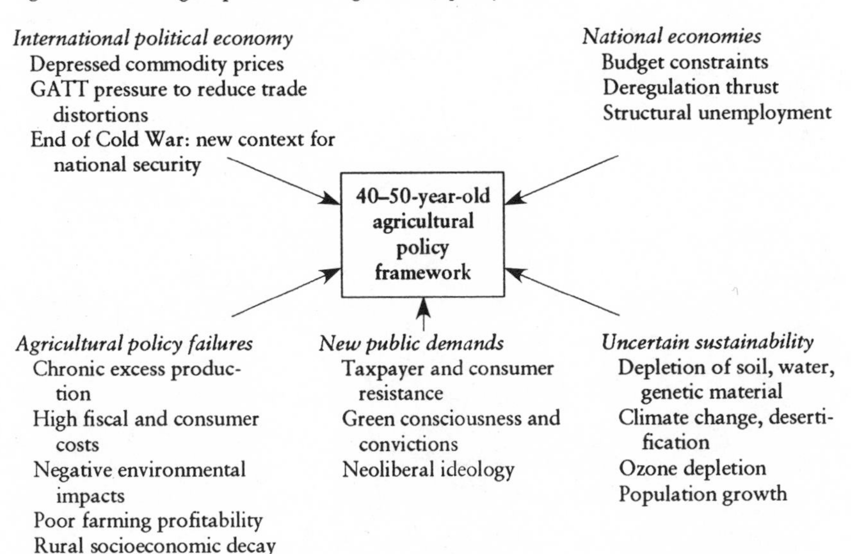

state, and the Punta del Este signatories—obsolete policies were not consigned to the dustbin of history. Indeed, the modest short-term success of measures to cut commodity surpluses and budget burdens may itself have diminished the sense of crisis, thus slowing the political momentum for more sweeping change.

In this environment of much talk and relatively little action, Sweden has moved ahead to enact and implement a wide-ranging reform of food policy: a process that might prove instructive—perhaps even mildly inspiring—to other nations. Swedish politicians, after all, claim that the new policy will improve economic efficiency and consumers' well-being while simultaneously maintaining food security, protecting the environment, encouraging energy self-reliance, and enhancing agriculture's nonproduction values—all this without imposing punitive burdens on farmers (Fornling 1990). If the politicians are even half right about this magic bullet, there may be something to learn from Sweden.

3

The Swedish Food System, Past and Present

Sweden lies in far northern Europe, yet most of it has a temperate climate because of the Gulf Stream's moderating influence. Natural conditions for agriculture are quite diverse, however. The Norrland region makes up two-thirds of the country, and in the far north a growing season of only 140 days restricts crop production essentially to potatoes, grass ley, and pasture. Farther south, but still in the Norrland, oats and barley enter rotations as fodder crops. Dairy and meat are the Norrland's comparative advantages, and they survive on the basis of special state supports, despite a widening absolute cost disadvantage vis-à-vis southern Sweden. (In international competition, northern Sweden's long winters and related high costs for shelter and winter feed would make even milk and beef nonviable.) The Norrland's limited arable land, just 12 percent of the national total, is typically found in clusters of small fields scattered through a vast forest landscape. Indeed, the typical northern farm has several times as much forest as arable acreage, and without woodlot operations few northern farms could provide their owners full-time employment or a living income. Most of the cultivable soils are relatively poor silt, moraine, or organic (peat) types located in river valleys and on the Bothnian Coast.

The southern third of Sweden—Götaland and Svealand—has 88 percent of the arable land and an even larger proportion of the full-time commercial farm operations. Just over half of all arable land and 60 percent of agricultural production are located in rich alluvial clay plains that extend from the south coast to a bit north of Stockholm. Here the yields of small grains, oilseeds, and sugar beets[1] are compa-

[1] Sugar beets are economically viable only in the far south, where the growing season reaches 240 days.

rable with those of most of continental Europe. As a rule, these stable, fertile soils are subject to little erosion. A cool and relatively dry climate keeps pest and pathogen problems to a comparatively low level. Southern Sweden's open plains are interspersed with large areas of forested uplands and mixed farm-forest zones where cultivated and pasture fields intermingle with small woodlots (Figure 3.1). Throughout Sweden, open crop and grazing land contributes to a visually and biologically rich mosaic of fields, forests, wetlands, and lakes.

3.1 Evolving Agrarian Land Use

After the end of the last ice age, roughly ten thousand years ago, Sweden was gradually revegetated. Bands of hunter-gatherers followed the receding ice edge, and in the Neolithic period they began to keep domesticated livestock. The oldest traces of cultivation are from five thousand to six thousand years ago. In early historical times, large parts of the present land mass were still covered by the sea; thus much of today's arable land was not previously forested, but consisted of marshy meadows that were grazed and later cultivated as they dried out with the rise of the land or were artificially drained in recent centuries.

When the climate became colder around 500 B.C., herders adapted by producing fodder for the winter months. A pattern of cultivation developed that in its main principles persisted into the nineteenth century. Tilled fields lay near village settlements. They were few and small, often fallowed, and fertilized by manure collected from the sheds where cows were kept in winter. Hay was gathered from meadows surrounding the tilled fields. The pattern of nutrient flow, from grass to manure to cultivated crops, gave rise to the adage that "the meadow is mother to the field." The ratio of harvested meadows to tilled fields required to sustain fertility varied between 5:1 and 20:1. Cattle were seasonally grazed in open forest land and later in fenced land with scattered trees *(hagmark)* beyond the fields and meadows. Over its two-thousand-year history, this stable land-use pattern, with many site-specific variations, became closely associated with rich cultural traditions as well as rich biotopes (Figure 3.2).

Agriculture and the landscape changed radically in the nineteenth century, as declining death rates accelerated population growth to over 1 percent per annum and put added pressure on the land. Wholesale industrialization came relatively late to Sweden, and two-thirds of the

Figure 3.1. Geographic distribution of agricultural activity

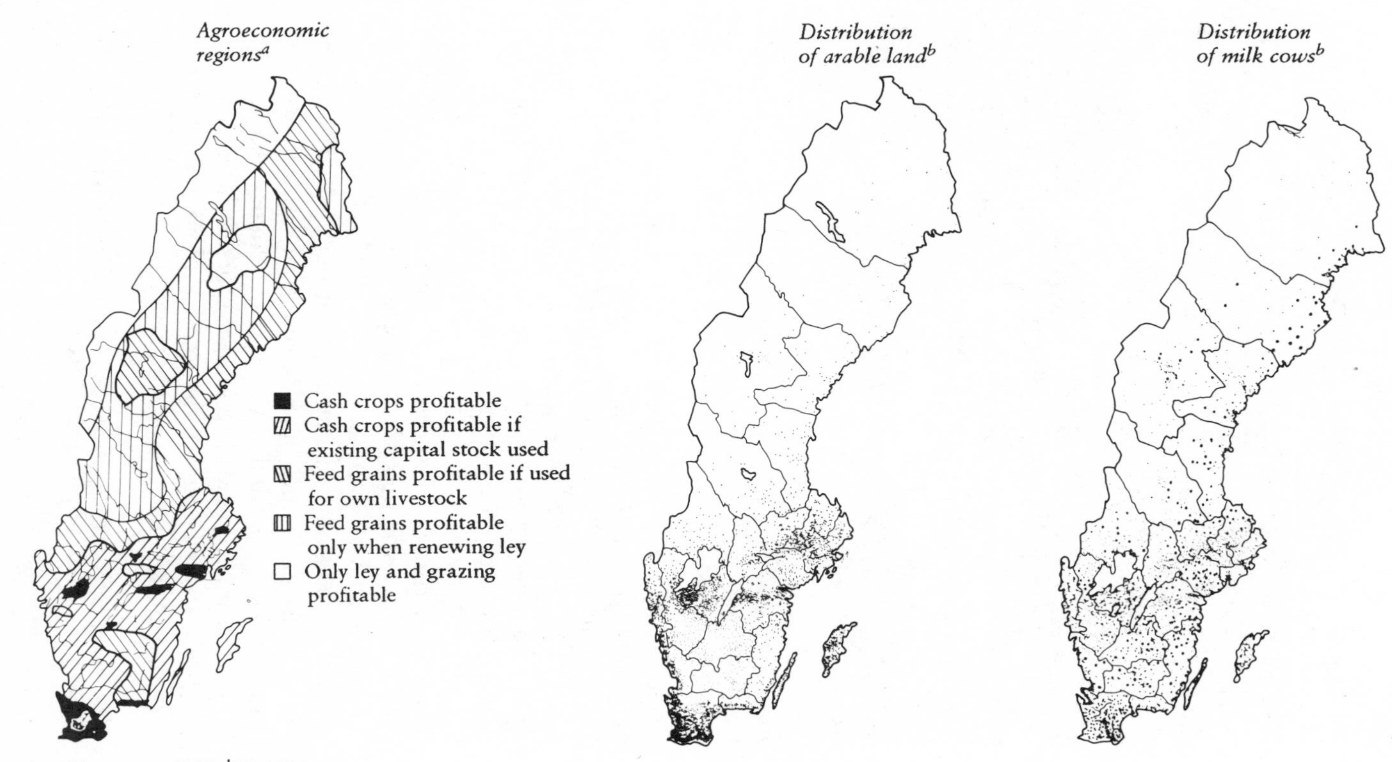

Sources: [a]Jonasson 1989, [b]SS 1990a.

Figure 3.2. Small-scale agriculture in a mixed farm-forest region (photo by Eddie Granlund, courtesy of *Naturbild*)

economically active population was still in agriculture in 1880. State-mandated land reform, one element in a plan to rationalize the "mother industry," burst the clustered villages into scattered family farms. The expanded acreage of cash crops and the introduction of cultivated fodder crops, especially nitrogen-fixing clover, into rotations dramatically altered cropping systems and the landscape. Figure 3.3a suggests the transition by which meadows, permanent pasture, and fallow were converted to rotational pasture and tilled land. Economically marginal pastureland also began to be reforested after 1900, as was arable land after 1930.

Food production and marketed surpluses increased considerably, but growing population and depletion of nutrients in meadow soils made starvation a familiar guest at the dinner table during much of the nineteenth century. Land hunger spurred the reclamation of virtually every arable hectare in the south,[2] colonization of marginal lands in the north, and large-scale emigration.

[2] State investment in hydrological research and reclamation made it possible to convert shallow lakes and marshlands into fertile fields.

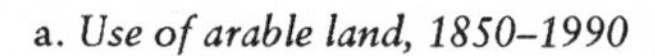

Figure 3.3. Past and present land-use patterns

a. *Use of arable land, 1850–1990*

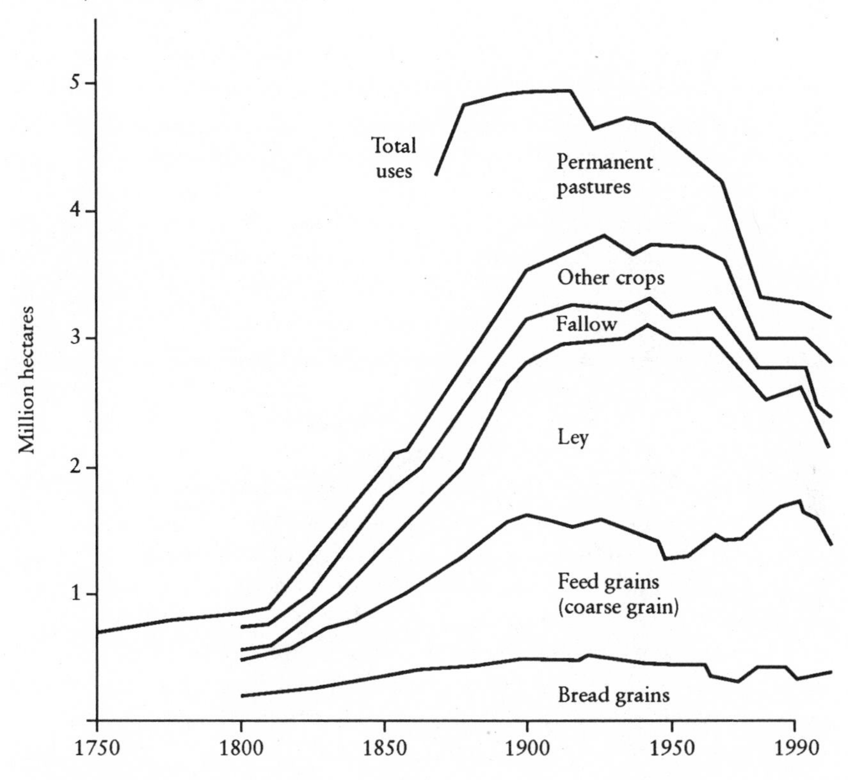

Source: SS statistics.

b. *Use of arable land, 1989*

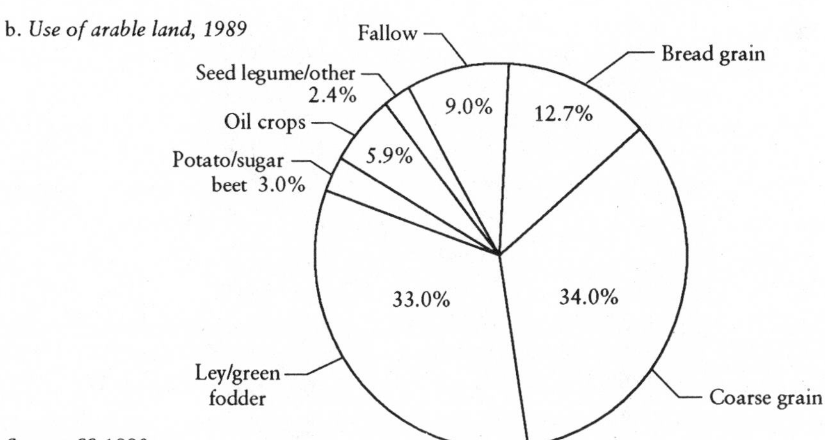

Source: SS 1990a.

3.2 Changing Resource Use in the Postwar Period

Over the past half century, Swedish agricultural evolution has been shaped by forces and characterized by processes generically similar to those in other advanced capitalist nations. Grain acreage increased steadily from 1800 until well into the present century; it rose yet again from 1950 until the latter 1980s when an acreage set-aside was imposed (Figure 3.3a). Essentially all the twentieth-century expansion has been in feed grains, reflecting the minimal demand growth for bread grains, the increasing predominance of livestock products in consumption, and the labor-saving advantages of growing and feeding grain instead of grass production in the agricultural economy. Grain's relative share of cultivated land increased steadily to reach a peak of 54 percent in the 1980s. Hay and silage declined simultaneously and today occupy only one-third of arable land (SS 1991a).

Cultivated acreage peaked at about 4 million hectares before World War II, declining gradually to 3 million hectares in the 1950s and 1960s. Losses were concentrated on marginal lands in the north and forest regions that were withdrawn from production as new mechanical and chemical technology amplified their competitive disadvantage and as farm workers departed for higher-income occupations. About one-third of arable land in the north was abandoned or afforested between 1950 and 1989. Although the labor exodus has continued to the present, a mix of labor-saving mechanization and policy measures—especially price supports, subsidies to disadvantaged regions, and land-use regulations—kept aggregate acreage nearly constant at 2.9 million hectares after 1970. We explain in Chapter 8 why the 1990 food policy reform is likely to cause yet another major reduction in cultivated acreage.

Analyzing the dynamics of U.S. agriculture since the Great Depression, economist Willard Cochrane (1979) described three overlapping revolutions: commercial specialization, mechanization, and biological-chemical innovation. Swedish agriculture has been transformed along analogous lines. Continuous farm productivity increases were driven by new technology. Since 1950, for example, milk yield per cow has doubled and wheat yields per hectare have increased by 120 percent. The dramatic rise in total factor productivity, or output per combined unit of inputs, is readily inferred from Figure 3.4.[3] Total factor productivity grew especially fast, roughly 4 percent per year, between 1960 and 1975.

[3] Note that these productivity estimates take no account of the social and environmental costs of new production methods and greater input intensity.

Figure 3.4. Trends in Swedish farm output and inputs (1950/51=100)

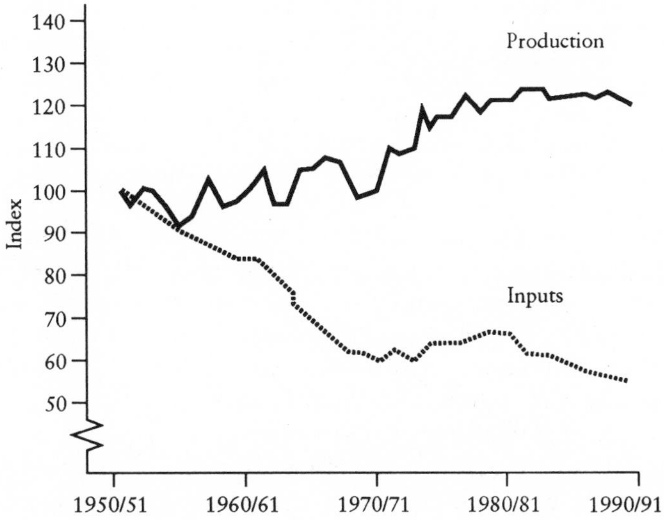

Source: Hjelm 1991.

For most of the postwar period, the rising cost of labor relative to machinery and fossil fuels was a powerful force inducing substitution of capital for labor (Table 3.1). These price trends also strengthened the economic incentive to expand capital-intensive farm enterprises, such as confinement hog operations, and contract labor–intensive ones, such as grass-fed beef. In the long run, industrial inputs have been substitutes for land as well as labor. Four decades of investment in machinery and structures raised the capital stock by nearly 400 percent in real terms.[4] This together with the laborsaving bias embod-

Table 3.1. Swedish farm input price trends, 1945–1989 (annual percentage change [1945 = 100])

Input	1945–66	1967–76	1977–84	1985–89
Labor	7.8	2.7	1.3	6.6
Machinery	0.7	0.9	1.4	6.2
Fuel and lubricants	1.2	1.5	4.1	−4.0
Fertilizer	0.3	0.8	2.4	0.4

Source: Hjelm 1991.

[4] The value of farm buildings increased by only 12 percent during this period, as the sharp decline in horses and the more gradual decline in cows reduced the need for stable space.

Table 3.2. Factor cost shares in Swedish agriculture (percentage of total cost)

	1950	1989
Labor	59	34
Capital and land	21	41
Purchased inputs[a]	20	25

Source: Hjelm 1991.
[a]Fertilizer, biocides, fuel, imported feedstuffs, services, etc.

ied in new equipment raised capital's share of agricultural value added from 21 percent in 1950 to 41 percent in 1989. Over those four decades, labor's share dropped from 65 percent to 38 percent, while the share imputed to land increased moderately, from 14 percent to 21 percent. Capital inputs per work hour grew ninefold, roughly in pace with the economy as a whole (Uhlin 1989:25).[5]

Purchases of variable inputs and services as well as fixed capital have grown in relative importance (Table 3.2). Application of nitrogen fertilizer, for example, more than tripled, from 68,000 metric tons in 1950 to 230,000 in 1989. Land-augmenting biological and chemical techniques have been the keys to agricultural intensification and yield increases.

The weighted index of total farm inputs (Figure 3.4) decreased by 40 percent between 1950 and 1970 as labor was squeezed out of agriculture. Thereafter aggregate input use rose mildly in the 1970s, under the influence of production shortages and productionist policy.[6] A reversal of market conditions and policy emphasis then renewed the downward trend in the 1980s. For the entire 1950–90 period, cultivated land contracted by 20 percent and labor inputs by fully 86 percent, but a doubling of total productivity boosted real output by 25 percent (Hjelm 1991).

3.3 Evolving Farm Structure

Sweden's agricultural labor redundancy turned to labor scarcity during World War II, as it did in many other countries. After the war, the inducement to mechanize and more generally to rationalize farming

[5] Uhlin uses neoclassical marginal productivity analysis to estimate factor shares of farm value added. Since farm price supports have been capitalized into land values, the share of value added attributed to land may have an upward bias.

[6] The exodus of workers from agriculture was also retarded in the latter 1970s by high nonfarm unemployment.

Table 3.3. Size distribution of arable land holdings, 1989

Holding size (ha)	Number	Percentage of	
		Holdings	Land
2.1–10.0	34,955	35	7.3
10.1–20.0	21,349	21	11.1
20.1–50.0	27,080	27	30.9
50.1–100.0	11,308	11	27.1
Over 100.0	3,861	4	23.5
Total	98,553	98	99.9

Source: SS 1990a.

methods was strengthened by two factors: the rapid growth of non-farm wages and employment, and the large nonwage component of farm labor costs.[7] High taxes on income, relative to taxes on capital assets, also reinforced farmers' incentive to plow back revenues into investments in machinery, structures, and land. Swedish experience thus confirmed the old saw that farmers tend to live poor and die rich. High labor costs, including the opportunity cost of farm families' own time, contributed to several trends in addition to mechanization: a disruption of intergenerational farm transfers, an increase in part-time farming, and a deterrence of capitalist relations in agriculture. A policy commitment to income parity between farm operators and skilled industrial workers, first made in 1947, slowed the labor exodus. Still, agriculture's share of the economically active population fell from one-fifth at the end of the war to only 3 percent by the 1980s.[8] The departure of young people steadily raised farmers' average age, so that one-third are now over sixty years old.

The number of farms fell steadily from nearly 300,000 in 1950 to 155,000 in 1970 and 97,000 in 1990 (Table 3.3). For most of the period, it was primarily farms smaller than 10 hectares that "disappeared" (declining from 245,000 in 1950 to 35,000 in 1989). With land consolidation, the average holding grew from 12 hectares of ar-

[7] The bias against using hired labor in Swedish agriculture is stronger than in many other countries because labor costs are increased by tough occupational safety regulations and mandatory employer contributions to pensions, paid vacations, sick leave, and accident insurance. These contributions, which cover nearly all seasonal and part-time workers and not just full-time employees, make total labor cost almost 50 percent higher than wages. Income taxes are also much higher than taxes on working capital.

[8] As more and more farms have become part-time, leisure, or retirement operations, agriculture's share of national labor input—about 1.8 percent—has fallen below its 3 percent share of the labor force.

able land in 1950 to just under 30 hectares, so that today nearly 40 percent of farms and 60 percent of arable land are in middle-sized holdings of 20 to 100 hectares. The share in very large holdings has steadily increased, while small holdings are no longer contracting significantly, since most of their owners rely primarily on nonfarm income. The growth of what active farmers disparagingly label "sidewalk farming" is reflected in the expansion of leasehold arrangements from a negligible fraction of land in the 1930s to 43 percent today.[9] Individual and family operators still cultivate 92 percent of arable land, while nonfamily corporations manage a small but fast-growing acreage (79,000 ha in 1986 and 128,000 in 1990) (SS 1991a).[10]

Table 3.4 summarizes many features of farm structural change, concentration of capital, and managerial/technological rationalization over the past four decades. The sharp drop in value added per farm and per work hour in 1990 compared to 1970 reflects the farm crisis of recent years.

Structural change has meant far-reaching specialization.[11] While dairy production remains geographically dispersed (see Figure 3.1), milk cows are no longer a feature of every farm. Herds of more than fifty cows have become common. Seven percent of Sweden's fifteen thousand hog farmers, with more than five hundred pigs each, produce half of total output. Poultry production is even more concentrated, with five-sixths of the flock in stocks larger than five thousand fowls (SS 1990a). Large-scale livestock confinement operations are concentrated in the far south, where production costs for feed grain are lowest, winters are relatively short and mild, and farms have historically been larger and better capitalized.

[9] Farm ownership attracts nonfarmers partly for recreational benefits and family-historic associations. Until the tax reform of 1990, it also had financial advantages: preferential tax treatment and the capitalization of high farm prices into land values and rents. Agricultural and tax reforms discussed in Chapters 8 and 9 have drastically reduced these incentives. The large amount of lease land also results from restrictive legislation regarding acquisition of farm real estate. Until very recently, owners who attempted to sell farmland usually faced price ceilings. Once out of the sector, they found it nearly impossible to purchase a new farmstead.

[10] A growing number of family operations have been incorporated for tax and inheritance reasons. Apart from experimental farms, the state is not involved in farm operations.

[11] Woodlot management, historically a crucial part of the typical farm's economy, has also become more specialized. In 1950, 8 million hectares of forest (35% of the national total) were managed in conjunction with agriculture. Today only 4 million ha are linked to farm holdings, and half of that is in part-time and hobby farms (Eriksson 1989).

Table 3.4. Changing farm structure, specialization, and productivity, 1950–1990

	1950	1970	1990
Farm characteristics			
Number of farms[a]	290,000	155,000	96,600
Arable land per holding[a]	12	20	29
Farms over 100 arable ha	2,300	2,600	4,003
Percentage of farms with			
Dairy production	92	62	27
Pigs	48	37	15
Number of milk cows (1,000s)	1,590	750	580
Livestock per farm[b]			
Cows	6	8	22
Pigs	7	36	158
Laying hens	39	78	480
Tractors per 100 ha	1.6	5.9	6.4
Nitrogen fertilizer (10^3 mt)	68	226	230
Biocides (mt)[c]	6,865	8,556	5,500
Productivity indicators			
Gross sales/Farm (1,000 SEK)[d]	NA	$31,750	$51,590
Value added/Farm (1,000 SEK)[d,e]	NA	$8,250	$6,980
Value added/Work hour (SEK)[d,e]	NA	$9.52	$2.70
Wheat yield (kg/ha)	2,440	3,750	6,560
Milk yield/Cow (kg/yr)	2,910	4,180	6,880

Sources: Less 1976; SS *Yearbook of Agricultural Statistics,* various years.
Note: NA = not available.
[a]Holdings over 2 ha.
[b]Includes only farms that raise the animal commercially.
[c]Active ingredients of insecticides, herbicides, and fungicides. Concentration of active ingredients and impact per gram have increased over time.
[d]Measured in constant 1990 kronor, then converted at the 1990 exchange rate of SEK 6.3 = US$1.
[e]Value added = total farm sales − expenditure for production inputs.

As previously indicated, farms run by family labor remain overwhelmingly predominant. Nonfamily corporations own just 3 percent of arable land, and capitalist operations, defined simply as those where wage earners perform more labor than the owners, are almost entirely limited to factorylike pork, egg, and poultry operations. The seventy-five hundred permanent farm employees contribute just 6 percent of aggregate farm work time. In all, wage labor including seasonal and part-time work is about 10 percent of the total (LSR 1989:49, SS 1990a, Swedish Institute 1988).

As in most industrial nations, farm structure is becoming bipolarized, with the absolute number of farms declining in all but the largest size category and the proportion of very small operations also increas-

Figure 3.5. Net farm household receipts from different sources, 1989 (by size of land holding)

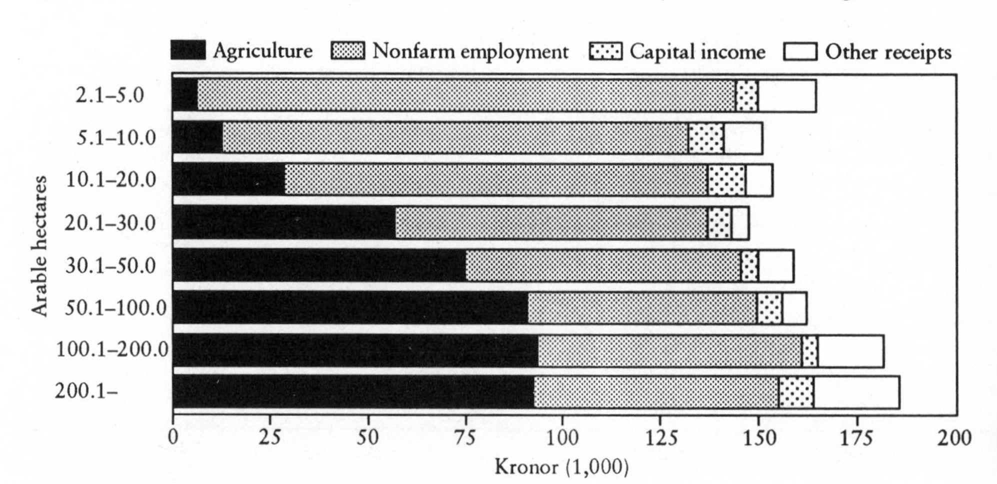

Source: SS 1991.
Note: Includes farm operator(s) and spouse(s).

ing. For reasons we explain in Chapter 7, the subgroup of organic farmers expanded at an extremely rapid rate in the latter 1980s, but they remain economically marginal.[12] For farm operators as a group, more than two-thirds of total household income now comes from non-farm sources. Figure 3.5 summarizes the relationships among holding size, total household income, and income sources.

Analyzing farm structure in terms of labor inputs, we see that only 39 percent of farms had at least one full-time worker in 1985 (full time is defined as 1,600 hours per year), yet these farms cultivated 68 percent of the land and accounted for 80 percent of farm sales (90% of livestock products). Slightly over half these farms are in the plains of south and central Sweden. Table 3.5 shows how labor time correlates with the distribution of land holdings, farm sales, and farm income.

Agriculture is a small economic sector that contributes only 1.5 percent of Swedish GNP and continues to shrink. Primary farm production has been subsumed within the larger agroindustrial system and now accounts for just 25 percent of the food system's employment

[12] In 1986, 520 farms (4,950 ha) were certified as organic or biodynamic; by 1990, the figure was 1,810 farms (28,900 ha) (SS 1991b).

Table 3.5. Percent shares of farm employment and production, 1985

	Work hours per farm				
	Less than 800	800–1599	1600–3999	4000 or more	Total
umber of farms	38,000	22,000	32,000	8,000	100,000
are of					
Farm employment hours	10	15	49	26	179.8 million
Arable land	14	18	44	24	2,728,000 ha
Total farm sales	7	13	47	33	SEK 31.8 billion
Net agricultural income	1	15	62	24	SEK 3.0 billion

Source: Hjelm 1991.

and 20 percent of its value added. Continuous technological change and intensive input use pushed gross output per hectare to US$1,107 in 1986/87, measured at internal support prices (the figure is less than half as great as world market prices). Although demand trends and policies caused a slight decline in the share of livestock products, they still account for 80 percent of gross farm sales, with dairy products alone representing 40 percent (Hjelm 1991, NAMB 1988b, SS 1990a).

3.4 A Comparative Sketch

To put contemporary Swedish agriculture into context for readers, we compare several of its basic features with those of two nations whose agriculture is better known on the world scene: France and the United States. The categories in Table 3.6 highlight several patterns of similarity and difference. First, the Swedish agricultural sector is small in absolute terms and plays only a minor role in international commodity markets. Second, Sweden is intermediate between France and the United States in typical farm size, resource intensity, and crop yields, though in general it more closely resembles France. An important exception is Sweden's intensive dairy production. France's superior yield in cereals stems from a combination of more favorable climate and fertilizer intensity; the low U.S. yields are linked to its land abundance and lower fertilizer applications. Third, the average size of French and Swedish livestock operations is far smaller than in the United States, although both European nations have fast-growing industrialized confinement subsectors. Pork accounts for 70 percent of

Table 3.6. Comparative agricultural features in Sweden, France, and the United States

Feature	Sweden (1990)	France (1990)	USA (1988)
Indicators of sector and farm size			
Total arable land (10^6 ha)	2.8	29.4	189.9
Number of farms (10^3)[a]	97	1017	2159
Arable land per farm (ha)	29	28	88
Total milk cows (10^6)	0.6	5.7	10.3
Cows per dairy operation	23	21	52[b]
Percentage of labor force in agriculture	2.1	8.5[c]	2.6
Indicators of resource intensity			
Tractors/100 ha arable land	6.3[d]	5.2	2.5
Nitrogen fertilizer (kg/arable ha)	80	95	55
Insecticide (g active ingredient/ha)	14	263	221[b]
Herbicide (g active ingredient/ha)	649	1273	1320[b]
Indicators of yield			
Average yields (mt/ha)			
Winter wheat	6.6	6.6	2.7
Barley	4.6	5.7	2.8
Total production (10^6 mt)			
Bread grains	2.6	32.6	57.8
Coarse grains	3.7	20.3	215.2
Average yield per dairy cow (mt/year)	6.9	4.9	6.2
Share of world exports, 1984/85–1985/86 average (%)			
Wheat	0.5[e]	16	30
Pigmeat	2.6[e]	NA	2.5

Sources: NAMB 1991; OECD 1988a, 1991b; SCEES 1991; SS 1991a; UIPP 1991; USDA 1989, 1990a, 1990b.

Note: NA = not available.

[a]In Sweden farms are defined as holdings over 2 ha; the French threshold is 1 ha. Farms in the USA must have at least $1,000 in annual sales.

[b]1990.

[c]1988.

[d]1986.

[e]1982–86 average.

Swedish meat production, compared to only 27 percent in France and 23 percent in the United States. In contrast, Sweden's poultry sector is exceptionally small. Fourth, Sweden's comparatively low use of biocides, partly due to green policies (see Chapter 7), is largely attributable to its long, cold winters, cool growing season, and small acreage of susceptible crops. Finally, Sweden is quite similar to the United States in the minimal proportion of its workforce engaged in agriculture. These countries' socioeconomic systems are less agrarian than that of France.

Table 3.7. Swedish self-sufficiency ratios and world trade shares

Commodity	Self-sufficiency ratio 1979/80	Self-sufficiency ratio 1985/86	Share of world exports, 1985/86 (%)
Wheat	131	154	0.5
Rye	107	126	1.5
Barley	107	126	1.3
Skim milk powder	128	204	0.6
Butter	119	122	0.6
Cheese	88	92	NA
Beef	102	110	0.4
Pigmeat	108	120	2.2
Poultry	102	100	0.1
Eggs	106	115	0.5
Sugar	93	95	NA
Oil cake	22	50	NA

Source: OECD 1988a:123–124.
Note: NA = not applicable.

3.5 Self-sufficiency and Trade: Sweden as a World Player

In 1987 Sweden's net imports of agricultural products were US$160 million, but if trade in farm inputs and processed foods is included, the net import bill was nearly ten times as large.[13] Sweden produces surpluses of many temperate commodities, however, and they grew sharply in the first half of the 1980s (Table 3.7). Even for imported commodities, such as cheese, sugar, and protein feed, import dependence declined in these years. Table 3.7 supports two further conclusions: first, the policy of shifting surplus resources from livestock to grains worked poorly in the first half of the 1980s; second, Swedish export surpluses are a very small share of world trade, even though they have been a major domestic headache.

3.6 Organizational Density and Agroindustrial Linkages

Swedish farmers are not atomistic individualists. Throughout this volume we stress farmers' cohesive economic and political organiza-

[13] Almost all fertilizers and biocides as well as 70 percent of farm equipment are imported. This dependency has had a great impact on the national food security debate, as we discuss in Chapters 4, 5, and 8.

Table 3.8. Market share of Swedish farmer cooperatives, 1984 (percentage)

Inputs and farm finance		Farm commodities	
Fertilizer and lime	80	Dairy products	100
Purchased animal feed	86	Slaughter	78
Machinery	40	Flour milling	80
Insurance	85		
Commercial credit	86		

Source: Bolin et al. 1986:20, 39.

tion. Sweden's first dairy cooperatives appeared in the 1880s, and a national cooperative association was formed in 1905. The National Farmers' Union dates to 1929.[14] Today Swedish farmers are highly organized, by locale and by commodity. More than 70 percent of farm operators (83% of full-timers and 56% of part-timers) belong to the Swedish Farmers' Federation (Lantbrukarnas Riksförbund, LRF). LRF members' involvement in organizational activities is broad but shallow. On average they belong to seven LRF subsidiaries, ranging from input and commodity marketing co-ops to credit unions and local clubs, but fewer than half are directly involved in organizational activities in any given year (Micheletti 1990).

In recent years, the widening economic gap between full-time commercial farms and others, and conflicting positions on environmental issues, have prompted the formation of new organizations to represent small and part-time farmers and organic farmers. Some of these groups openly oppose the LRF on economic and political issues, and at one time they appeared to threaten its political-economic hegemony. Judged by sheer numbers or by shares of acreage and output, however, the insurgent organizations remain fairly marginal. When the debate over agricultural policy reform heated up in the latter 1980s, the LRF commanded most attention from politicians and the media.

Farmer-owned cooperatives, as shown in Table 3.8, have had a near monopoly in the markets for farm financial services, some inputs, and primary food processing. They owned 49 percent of Swedish food processing capacity in 1981.

Today few farmers are directly involved in managing cooperatively owned ventures, yet their equity stake in the co-ops is one strand in a

[14] In Chapter 5 we describe the process by which the state more or less forced milk producers to join a dairy cartel in order to participate in corporatist price setting during the 1930s farm crisis.

web of common interests tying Sweden's 97,000 farmers to an even larger number of wage earners in farm input and food processing industries. These well-organized interests have been key actors in the food sector's negotiated economy.

3.7 Summary

We have documented the most prominent structural features and evolutionary tendencies in Swedish agriculture. Many are broadly similar in kind to patterns in other advanced industrial nations: growing concentration of land and capital, loss of farmland, increased specialization, rising yields, laborsaving technical change, polarization of farm structure, a high degree of farmer organization, and the persistence of family-based farming. In the 1980s, chronic production surpluses and depressed farm profitability also became familiar themes in the Swedish agricultural story.

4

Swedish Agricultural Policy: Achievements and Contradictions

The evolution of Swedish agriculture shaped and was shaped by agricultural policy. Through shifting goals, priorities, processes, and instruments, this policy has evolved from the 1930s, when a distinct sectoral agenda first took shape, until the late 1980s, when the old regime was on the verge of collapse. How effective have these policy measures been in achieving their stated ends over the course of a half century? This is a daunting question because of the complex goal set, conflicts among some objectives, the multiple policy instruments, and the powerful nonagriculture forces that have influenced outcomes. Despite these impediments to interpretation, we find evidence of deteriorating goal attainment, which set the stage for policy reform.

In Chapter 5 we ground our analysis of Sweden's agricultural policy regime in a theory of the state and economy. Gøsta Esping-Andersen (1990) has shown that the modern state system in northern Europe's social democracies developed in historical circumstances distinct from those of the "liberal" and "conservative" states found in other industrial capitalist societies, particularly regarding the formation of political class coalitions. Swedish agricultural policy, specifically, originated in a class alliance between yeoman farmers and wage earners, forged in the economic crisis of the 1930s.

The agricultural policy regime, with its dense legal-institutional matrix, is construed as one moment in what some Scandinavian social scientists call "the negotiated economy." It has three fundamental features. First, resource allocation and income distribution result from a dialectical interplay between negotiation procedures and three other institutions: democracy (voting procedures), bureaucracy (administrative rules), and markets. Second, the state plays two often contradictory roles: it articulates and represents the public interest and it mediates conflicting private interests.[1] Third, private interest organizations' long-term involvement in negotiated relationships "tames" them to some extent, tempering their pursuit of self-interest with a concern for broader needs and interests (Micheletti 1990, Nielsen and Pedersen 1990).

We hypothesize that agricultural policy has been determined—or overdetermined—by the interplay of many forces, primary among which are the following: trends and conditions internal to the food production and distribution systems; contradictions in policy instruments; exogenous national and international economic conditions; the exigencies of coalition building in a multiparty democracy, interest group competition, and political culture, by which we mean widely shared values, meanings, and ethical norms.

The Swedish agricultural policy regime comprises five core processes:

- Major farm legislation, normally at intervals of five to ten years
- Detailed legislation to flesh out or adjust policy directives
- Detailed regulatory design, including annual price setting for farm products
- Agricultural budget setting within the larger fiscal framework
- Policy execution by the state's administrative agencies.

Since the 1940s, the process of designing major policies has followed a sequence of six phases. First, the government issues a directive and names a special commission to analyze the current situation, evaluate current policy, and recommend modifications. Second, the commission submits its report, with recommendations backed by analysis. Third, the commission report is circulated and responses are solicited from the affected public authorities, private interest organizations, and other concerned groups (the "remiss round"). Fourth, the government presents its legislative proposal to the Riksdag (parliament). Fifth, the government bill is reviewed and typically modified by the Riksdag's permanent Agriculture Committee. Finally, legislation is voted on in plenary session. Public debate occurs at every stage of this process.

[1] A third contradictory aspect of the state is that its agents—political parties, politicians, and bureaucrats—pursue their own vested interests even when cast in "disinterested" roles. This proposition is central to the public choice theory of the state, which we assess in Chapter 5.

Two distinctive features of the Swedish process are citizens' high regard for the commission reports and de jure participation by agents of all the affected interests.

In following our description of the twists and turns of legislation over the decades, readers should not lose sight of two central facts: first, the primary policy instruments—domestic market regulations and import protection—have remained in place since the 1930s; second, the regulatory apparatus needed to implement policy objectives has grown increasingly complex.[2]

4.1 Fifty Years of Agricultural Policy: Stability and Adaptation

Before 1930

Sweden had no distinct agricultural policy before the 1930s, yet numerous policies affected the farm sector and several can be considered precedents for modern farm policy. Taxation of rural society financed much of the state's expenditure. Successive land reforms starting in the eighteenth century consolidated fragmented parcels and established permanent ownership titles. These reforms encouraged the development of an agriculture based mainly on family farms. This national commitment was backed by universal public education dating from the 1840s and the creation of an extension service in the 1870s. After 1903, corporations were forbidden to acquire farm and forest land. Beginning in 1888, customs duties were used to protect farmers from cheap foreign grain—at the expense of consumers. Before World War I, agricultural production was promoted by large-scale drainage projects to reclaim fertile wetlands. A combination of serious food shortages and rampant price inflation during the Great War established popular tolerance for costly national food security measures (Heckscher 1915, Hedlund and Lundahl 1985).

In the decade following World War I, Sweden briefly reverted to nearly free trade in farm commodities. But with world market supply expanding and demand relatively stagnant, Swedish farmers, like their

[2] A sense of this complexity is conveyed by Ewa Rabinowicz: "The regulatory system in agriculture is very complicated. For example, almost 300 different variable [import] levies are determined for food products (excluding fish). For pork products alone, 54 different export subsidies (for different cuts) are calculated" (1991a:1–2).

counterparts in much of the industrial world, suffered recessionary conditions well before the financial and industrial crisis of 1929–30.

1930–1946: Interventionism and Corporatist Arrangements

In the 1930s, a distinct agricultural policy gradually took shape through a series of ad hoc measures. Sweden's economic depression and the political reaction to it were similar to those in many other industrial democracies. Protracted economy-wide crisis led to the replacement of an ineffectual conservative government with a left-leaning coalition. Farmers, who constituted nearly one-third of Sweden's working people and voters, faced economic devastation. For political as well as economic reasons, any government, left or right, had to respond to their plight.

In a historic compromise, the ascendant Social Democratic party (SAP) abandoned its commitment to free international trade in order to build a coalition government, with the Agrarian party as junior partner. The compromise, dubbed "the cow trade," required the Agrarian party to accept the SAP's prototype-Keynesian fiscal intervention for macroeconomic recovery in return for measures to raise and stabilize farm incomes.

Income maintenance was the policy's prime objective, though it was not yet expressed in the form of an explicit income target. Import protection and domestic price supports became the main policy tools. Social Democratic economic advisers understood the instrumental value of higher, more stable farm incomes, both as a demand stimulus and as a deterrent to the migration of impoverished farm workers to cities, where unemployment was already high. Nonetheless, these rationalizing intellectuals strongly opposed any permanent regime of protectionism and price supports. With prescience, Gunnar Myrdal argued in 1938 that "the peril our policy has to face and master [is] the *peril of overproduction*"; Karl Åmark cautioned against "emancipating production from demand so that responsibility for appropriate production rests upon the community" (Myrdal 1938, emphasis in original; Åmark 1942).

A National Agricultural Marketing Board (NAMB) was created in 1938 as the official price-setting agency. Building on corporatist institutional precedents in other economic sectors, NAMB negotiated farm prices with representatives of the national farm organizations. Since farmers were not yet organized at the national level, state-imposed pricing forced the consolidation of hundreds of primary marketing cooperatives into a national federation. This, then, was the unnatural birth of what was soon to become a powerful lobby.[3]

The reactive character of these policy measures should be stressed: "Interventions in agricultural markets were not a result of a carefully prepared strategy [and] regulations were introduced in an ad hoc manner when different commodities were hit by crisis." Thus, dairy producers were protected by a tax on margarine, grain output was limited by quotas, and surplus pork exports were financed by slaughter fees and a levy on protein feed (Rabinowicz 1991a:17–18). According to Hedlund and Lundahl (1985), the policy's only consistent feature was the priority given to farmers' economic interests.

The regulatory system was spared criticism by World War II, when the national security objective took precedence. Protectionism and still higher producer prices were justified by the urgent need to expand production of staple foods. With supplies tight after successive poor harvests in 1940 and 1941, the state ordered a diversion of resources from "luxury" livestock products toward staple crops. The state's "visible hand" effectively combined price incentives with food rationing to prevent a repetition of the First World War's hunger and inflation. As early as 1942, however, a special commission was assigned to recommend measures for more efficient attainment of the policy's ends once the war was over.

1947–1970: Policy Consolidation and Agricultural Rationalization

A comprehensive policy framework, the Magna Carta of Swedish agriculture, was legislated in 1947. It replaced the previous ad hoc regulations while retaining the core instruments of import protection and corporatist price setting. The 1942 commission, with its influential group of economists, took a social engineering approach. Agriculture was put in an overall social perspective and for the first time given a clear goal structure and consistent policy instruments. The policy consisted of two main categories of instruments: price and market regulations, and rationalization measures. The commission was heavily influenced by the wartime experience, and national food security emerged as the major justification for extending special treatment of the farm sector into the long period of peace and prosperity after 1945. The three goals set in the 1947 agricultural policy, albeit with

[3] Bo Rothstein (1991) has shown that the previous bourgeois government actually set the precedent for forced collectivization by mandating the formation of a milk producers' cartel. This was a crisis reaction on the eve of the 1932 election.

some changes in specification and relative priority, continued to be the rationale for special treatment of the farm sector down to the 1980s:

- The production goal: national food security
- The distribution goal: income parity and stability
- The efficiency goal: rationalization (to make high farm incomes compatible with reasonable consumer prices and overall economic growth).

Similar goals were common in western Europe following the war, and they appeared in the 1957 Treaty of Rome, which set the stage for the EC's Common Agricultural Policy.

Food security was interpreted as a capability of meeting 100 percent of the population's nutritional needs for up to three years in an international crisis and blockade.[4] This specification was modified over time, but food security was generally construed as peacetime self-sufficiency in staple foodstuffs other than sugar. This productionist strategy contrasts with the alternative of allowing peacetime production to respond to market forces, while ensuring sufficiency in a crisis by maintaining an adequate arable land base and stockpiling inputs and foodstuffs. Farm organizations lobbied vigorously for the productionist strategy, and its cost did not seem excessive in the formative postwar years, since commodity surpluses were minimal and international food prices were high.

The distributional goal, previously vague, was now expressed as an income parity formula. Farm prices should be set to give efficient family farms with ten to twenty hectares of arable land a net income roughly comparable to a skilled industrial worker's earnings. Changes in input costs, productivity, and nonfarm income were to be factored into the pricing formula. It soon became apparent, however, that prices targeted to reach an income standard for efficient, medium-scale farms would bankrupt large numbers of marginal producers, mostly small dairy farmers in geographically disadvantaged regions. Since viable agriculture in the Norrland and forest regions was considered essential, both to ensure national security and to protect the economic security of the least occupationally mobile smallholders, a regional milk price premium was instituted in the 1950s.

Dynamic efficiency in the farm sector was given priority as the principal way to raise farm income and meet national security goals with a minimum of hardship for consumers and taxpayers (food expendi-

[4]The basic nutritional objective was defined in terms of protein and energy needs (2,900 calories daily for adults).

ture still absorbed nearly one-third of disposable income). In addition, rising output per farm worker was seen as necessary to facilitate the shift of labor to the fast-growing industrial sector, thereby increasing national economic growth. Rationalization measures included assistance for land consolidation, subsidized investment credits, publicly funded research, farmer education, and extension services. Policies encouraged both external rationalization (consolidation of submarginal farms into larger, more efficient units) and internal rationalization (specialization and technical-managerial innovation).

It was understood that some marginal farmland would be converted to forest, with the proviso that sufficient land would be kept open to sustain a production base in disadvantaged regions. The rapid exodus of working-age farm people from agriculture was an explicit objective. The challenge was to find the right balance between income support and incentives to exit farming, to ensure "a well-ordered reduction of the agricultural population" rather than a "disordered catastrophic depopulation of rural areas" (MOA 1946:42).

Swedish agriculture changed dramatically in structure, productivity, and resource use between the 1947 farm legislation and the creation of the next agricultural policy commission in 1960. Rationalization proceeded apace, large numbers left for nonfarm occupations, the farmer–wage earner income gap narrowed, and yield increases brought Sweden to a position of self-sufficiency or surplus in most basic foods. In all these respects, the policy should be considered a success. But by 1960, domestic food prices were far above international levels and export subsidies had become the rule rather than the exception. Critics argued that Sweden had overshot the national security goal. Furthermore, although farm employment had plummeted from one-fourth to one-tenth of the labor force in barely fifteen years, there remained an acute labor shortage in the buoyant industrial and service sectors, evident in the need to recruit tens of thousands of "guest workers" from Finland, southern Europe, and Turkey during the 1960s economic boom. This shortage supported the argument that guaranteed farm income parity was blocking the flow of labor power from low to high productivity occupations and obstructing external rationalization. A related criticism voiced in the 1960s debate was that consumer welfare was being sacrificed unfairly to the economic interests of farmers and the agroindustrial sectors (Gulbrandsen and Lindbeck 1966, 1969).

In this atmosphere of criticism, the 1967 farm bill elevated the priority of efficiency and rationalization but without abandoning the national security and income goals. Greater emphasis was given to resolving the industrial labor shortage and, to a lesser extent, cutting the cost of export subsidies. Food security was redefined to mean only 80 percent peacetime self-sufficiency, and real farm prices were cut by 15 percent between 1967 and 1971 to discourage production and encourage resource outflow. The commitment to full income parity was thus undercut and labor outflow was indeed spurred—by a price-cost squeeze on marginal farms, by early pensions for dairy operators, and by an active labor market policy that invested heavily in retraining and resettling former farm workers. A separate action, consistent with the revisionist 1967 policy, was the creation of a delegation to represent consumers in farm price negotiations (Rabinowicz 1991a).

1971–1979: The Farmers' Golden Age

The late 1960s and early 1970s were years of rapid and partially unforeseen changes on the domestic and international agricultural scenes. The 1967 farm policy combined with escalating nonfarm wages to bring about a sharp contraction in farm labor input and some loss of farmland.[5] Farm work hours declined 30 percent between 1966 and 1971. The number of dairy farms fell 41 percent, and milk production dropped to its lowest level since the war. Most of the insolvent farms were in sparsely populated regions with poor soils and climate; the rapid loss of production capacity there was at odds with national security policy, and the abandonment of villages and loss of open landscape to "sterile" spruce forest began to evoke an adverse response from the public—and from smallholders, who took up the slogan "we won't leave!" Figure 4.1 illustrates this metamorphosis of the farm landscape, which most Swedes abhorred. These changes were intentional, but their pace and adverse consequences for farm people and rural Sweden soon provoked second thoughts.

In the inflationary macroeconomic environment of the early 1970s, consumer protests against high food prices broke out and the Social Democratic government responded by introducing retail subsidies. State expenditure on food subsidies peaked at US$750 million in 1980, equal to 20 percent of farm revenue from price-regulated commodities.

[5] Cultivated acreage declined by about 1 percent annually with losses concentrated in the Norrland (see Figure 3.3a).

Figure 4.1. Impact of afforestation on the farm landscape

a. *An agrarian landscape circa 1990*

b. *The same landscape with a thirty-year-old spruce plantation, 2020*

Source: Drake et al. 1991.

Subsidies had the dual effects of worsening the budget deficit and stimulating consumer demand, particularly for meat and dairy products. When growing demand confronted stagnant farm production in the mid-1970s, subsidized exports were replaced by rising imports, precisely when international food prices were moving sharply upward. Figure 4.2a illustrates the temporary catch-up between international and domestic prices in 1972–74.

The Swedish delegation to the 1974 World Food Conference in Rome joined others in pledging to increase agricultural investment, food reserves, and food shipments to the Third World. Back home, anxiety about food security and rural socioeconomic decay contributed to a "green wave" of support for farmers. This upwelling of sympathy with rural lifestyles and values was a part of a broader cultural tendency expressed in popular music, clothing fashions, and other ways. These expressions presaged a second green wave in the latter 1980s, which is the central focus of this book. A new agricultural commission formed in 1972 was naturally influenced by these events and public sentiments.

On the macroeconomic scene, the Social Democratic government was unable to engineer a quick solution to the "stagflation" of the mid-1970s. In 1976 a nonsocialist coalition government took power for the first time in more than forty years. Active farmers from the Center party (formerly the Agrarian party) were installed as prime minister and agricultural minister, so the stage was set for a new policy direction. Indeed, the 1977 agricultural legislation reasserted the priority of income parity and food security, while downgrading economic efficiency objectives. Not surprisingly, the principal means of promoting these goals was to raise farm prices and import levies. The budgetary cost of export subsidies was expected to be small; chronic Swedish surpluses and falling world prices were not foreseen. To the extent that a chronic surplus did arise, the intention was to adjust relative prices and so induce a shift of resources from livestock to grain.

Income parity between full-time family farms and nonfarm wage earners was reinstated as the top priority goal. The social preference for family farms was also reaffirmed, reflecting the Center party's ideological orientation and a popular reaction against the big, quasi-capitalist production units that were becoming more prominent in the livestock sector.

National food security was once again framed in terms of 100 per-

Figure 4.2. Domestic and international food price trends, 1960–1988

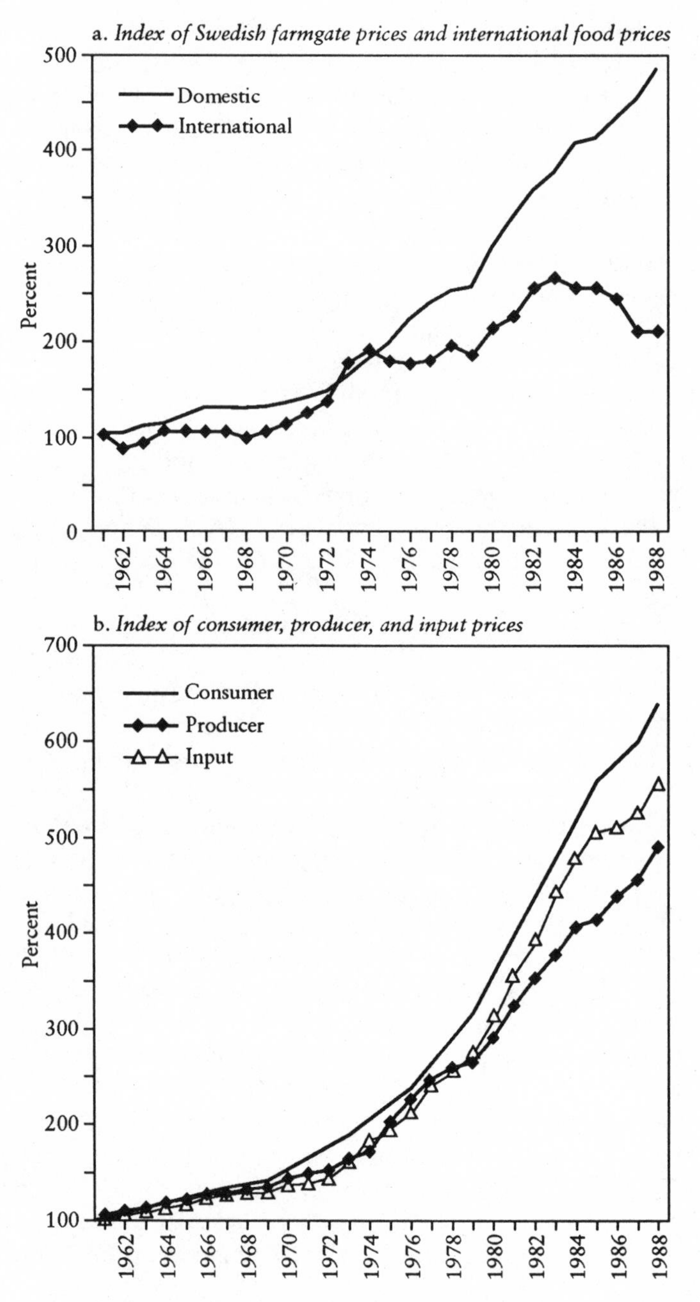

Source: Redrawn with permission from Rabinowicz 1991a.

cent peacetime self-sufficiency in staple foods (except sugar). This was justified in part as a gesture of international solidarity in a food-scarce world. Preparedness was again defined as maintenance of production capacity in every region, with a special emphasis on measures to offset the Norrland's long transport distance to input suppliers, processors, and markets. A special Norrland support and low-income supplements were intended to counteract steadily widening productivity gaps and income disparities between large and small farms and between the southcentral plains and other regions. These direct transfers supplemented the existing milk price premium to producers in economically disadvantaged regions. When the regional development goal for agriculture took on greater importance in the 1970s, farm income and food security goals became bound up with the broad problem of secular economic decline in sparsely populated regions. This emphasis was supported by green wave sentiments, but more important politically was the hope that a viable agriculture would help offset big employment losses occurring in the forest products, mining, and metals industries, which had been economic mainstays of Sweden's hinterland ever since the industrial revolution.

Food security was also the pretext for measures to maintain Sweden's remaining three million hectares of cultivated land. Trees already blanketed most of the land in economically marginal farming areas, and further afforestation was widely viewed as degrading nature, cultural-historical values, and recreational resources. A Land Acquisition Law (1979) strengthened the County Agricultural Boards' authority to intervene in land transactions to ensure that farmland would not be abandoned.[6]

Finally, the 1977 policy included an explicit consumer goal: "Consumers shall have access to food of good quality and at reasonable prices, and [they] must be able to be sure that production does not convey any risks regarding food quality or their health" (MOA 1977:17). The rhetoric about "reasonable prices" was vague, and in practice this goal was sacrificed to farm income and production objectives.[7] Professional economists, who had championed consumer wel-

[6] Intervention in the land market by local boards goes back to the early postwar period. Its basic weakness was that farmland protection could not guarantee agricultural profitability in marginal areas. The Land Acquisition law might, for example, prevent housing, commercial development, or forest planting, but it could not ensure continued agricultural use (Drake and Petrini 1985).

[7] In Chapter 5 we discuss the degree to which the official goals—international solidarity, food security, reasonable consumer prices, open landscapes, and so on—were genuine motives or pseudogoals covering for vested economic interests.

fare on earlier agricultural commissions, were conspicuously absent from the group that designed the 1977 policy. Indeed, the new policy, unlike its predecessors, was not publicly justified on the basis of economic logic (Thullberg 1983).

As the pro-agrarian policy came into effect, farm incomes rose substantially, farm investment increased, and farmland values soared. Most Swedish farmers enjoyed a brief spell of comparative prosperity.

The 1980s: A Policy Regime under Siege

Within a few years, the 1977 agricultural policy was rendered obsolete by internal contradictions and unanticipated international price trends. On the supply side, chronic excess production resulted from three forces: capital accumulation, increased fertilizer and chemical use induced by higher commodity prices, and the spread of yield-increasing biological and chemical technologies, exemplified by more potent pesticides and bovine embryo transplantation. The policy of shifting excess capacity from livestock to grain was not very successful (see Table 3.7), though Sweden did somewhat better than the EC in limiting surpluses of perishable, unmarketable dairy products.[8]

As production surpassed demand and international prices fell farther below domestic support levels, customs revenue was replaced by increasing export subventions. In 1986–87, subventions reached a peak of US$340 million, six times the level anticipated when the policy was set in 1977. The growing export subsidy bill coincided with intense fiscal stress, as the state's budget deficit surpassed 10 percent of GDP in the early 1980s. The nonsocialist coalition government, and then its Social Democratic successor, had to wrestle with an economic recession and a major industrial restructuring simultaneously. The urgent need to restrain government spending prompted a decision to phase out retail food subsidies. This action, and parity pricing during a period of escalating farm input prices, drove up retail food prices. The acceleration of food inflation and the widening gap between domestic and international prices are shown in Figure 4.2.[9]

When the Social Democrats returned to power in 1983, after a six-

[8] Sweden's comparative success in shifting surplus resources to grain is suggested by the following figures. Swedish wheat production was 165 percent of domestic consumption in 1984, compared to 145 percent for the EC. The self-sufficiency figures for dairy products are as follows: fluid milk—Sweden 127 percent, and EC, 177 percent; butter—Sweden, 123 percent and EC, 130 percent; cheese—Sweden, 93 percent, and EC, 118 percent (Vail 1993).

[9] Farm prices lag behind input prices in Figure 4.2b primarily because of rising farm productivity, but this lag also reflects declining farm profitability.

year hiatus, they quickly established a food policy commission. In a stark understatement, the minister of agriculture stressed in his directive to the commission that "the agricultural policy of 1977 does not work the way the Riksdag presupposed when the decision was taken" (MOA 1984:31). He emphasized that correcting past farm policy failures was necessary for achieving the nation's broad economic priorities of curbing inflation, shrinking the budget deficit, and restructuring the economy in a market-oriented direction. By calling the study group a food policy commission, the Social Democrats signaled the high priority of the consumer goal relative to the farm income goal (Andersson 1987).

There was too much disagreement among the political parties about the substance and pace of reform to make the 1985 agricultural legislation anything more than a temporary holding action. Nonetheless, the debate, the legislation, and a series of follow-up actions set the stage for more fundamental reform five years later.

Food security was reinstated as the central goal of state intervention, with farm income and consumer welfare accorded lower status. In the wake of growing public concern about fertilizer leaching, pesticide residues, and loss of farm landscape, environmental protection and nature conservation were made explicit farm policy goals for the first time. Food security was accorded highest priority more or less by default, since the Social Democrats refused to give top billing to income parity at a time when high farm prices were intensifying inflationary and fiscal problems. Moreover, many large farm owners in the plains already had higher incomes and far more wealth than most wage earners. The traditional notion of food security—a high level of peacetime production—drew increasing criticism, particularly from the powerful Ministry of Finance. Skepticism about the rationality of excess production was strengthened by waning East-West tensions, changes in modern warfare, and glutted world food markets. But with no national consensus on a new conception of food security or cost-effective measures to achieve it, import protection and price supports clung tenuously to life (Hedlund and Lundahl 1985).

The income parity formula was abandoned in 1985, signaling, first, the Riksdag's unwillingness to continue rubber stamping price agreements between LRF and the Agricultural Marketing Board and, second, its intention to strengthen the Consumer Delegation in price negotiations. It was also declared that farmers should bear the full cost of financing exports, which meant that at least indirectly they would have to face international price competition. In the interim, the state would finance 40 percent of grain export costs for five years. The growers' 60 percent share was to be financed through milling fees and levies on fertilizers and agricultural chemicals. Meat exports were financed by a surcharge on slaughter fees; and from 1985, dairy producers were subject to a quota system (more precisely, a two-price system, with above-quota production compensated at the depressed international price).[10] In sum, while nominal farm prices continued to rise, these ad hoc measures depressed real net farm prices and caused the classic cost-price squeeze on farm income (Figure 4.2b).

Price restraints and input levies failed to cut the grain surplus. On the basis of the cost-sharing formula adopted in 1985, the state's export cost burden was expected to be US$60 million/year. But in the new policy's very first year it was already $170 million, and in 1986/87 it peaked at $215 million (MOA 1987b:2). An emergency Small Grains Group was convened in 1986 to find a solution. It recommended a mix of short-term and permanent acreage set-asides, and by 1987 a voluntary program was in place to compensate growers for fallowing land or planting forest.[11] This program continued with modifications through the 1990 growing season. In 1988 and 1989, about 260,000 hectares, or 17 percent of grain land, was removed from crop production. By this expedient, grain exports were halved in 1989 and export subventions were cut by $100 million. The net budgetary saving, after compensating set-aside participants, was $35 million (NBA 1989c). Given the fundamental flaws of the protectionist regime, this was a cost-effective expedient. But few believed it was a sustainable solution.

4.2 Policy Evaluation: Goal Attainment and Cost-Effectiveness

The relative priority of food security, farm income, and efficiency has fluctuated over time, and new goals—regional development, consumer welfare, and environmental protection/nature conservation—

[10] Fertilizer had a 15 percent ad valorum levy, while pesticide levies were based on weight of active ingredient. An incidental objective was to encourage less intensive use of fertilizer and biocides. The 1984/85 slaughter fee on pork was 25 percent of carcass value. Several additional ad hoc measures were taken to reduce output. For example, a price premium was paid for veal to discourage raising steers, and early pensions for dairy farmers were reinstituted.

[11] Compensation was regionally differentiated according to average yields.

Figure 4.3. Evolution of Swedish agricultural policy priorities, 1880–1990

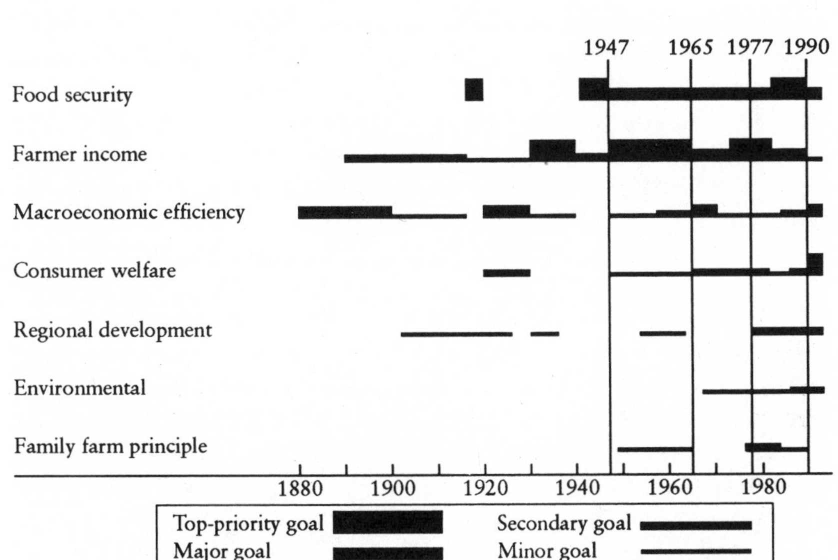

Source: Authors' interpretation of parliamentary decisions.

have been added. Citizens' collective willingness to pay for agricultural policy has risen and fallen several times since the 1930s, depending primarily on macroeconomic conditions. But whatever new measures were adopted, the core instruments—variable import levies, domestic market regulations, and corporatist price setting—remained in place. The trend has been toward more complex regulatory and incentive structures, driven by contradictions within the food production system as well as the policy's negative effects and its simple inadequacy to meet multiple objectives. Figure 4.3 shows the priority ranking among objectives, from 1880 to the late 1980s.

Ambiguity and analytic complexity are inevitable in policy evaluation, especially when exogenous forces besides agricultural policy directly and indirectly influence outcomes. Can agricultural rationalization measures be credited with a rapid and harmonious outflow of labor from farming after World War II, or were national labor market policies and high nonfarm wages the keys? Second, given multiple goals and multiple instruments, a single instrument may affect several objectives, some positively, others negatively, and goal conflicts may be unavoidable. Thus high support prices improve farm income but also stimulate excess production and cause greater pollution. Third, an instrument's effectiveness may deteriorate over time. In the food-scarce 1970s, the milk price premium to disadvantaged regions was a low-cost way to pursue national security and regional goals, but this ceased to be true in the 1980s when the milk surplus reached 27 percent. The following summary cannot account for all such complexities; it is necessarily impressionistic and reflects our viewpoint as economists.

The Production Goal: National Food Security

During World War II and much of the postwar era, maintaining capacity to provide adequate nutrition during a period of international hostility and blockade was a central justification for protectionism and price supports. In actuality, imports of food and farm inputs have never been obstructed, but that does not invalidate the collective desire for some form of "food insurance." Nevertheless, defining food security, setting concrete policy objectives, and choosing effective policy measures are fraught with ambiguity and uncertainty. The risks associated with low-probability crisis scenarios must be assessed and compared with the more-or-less known costs of preparing for them in alternative ways. Until 1985, Swedish farm policy was based on crisis scenarios that resembled shortened versions of World War II. The policy goal was to maintain a capacity to meet specific nutritional standards for a period of three years without recourse to international markets. This was translated into peacetime production targets for particular commodities and a goal of maintaining production capacity in regions disadvantaged by remoteness, thin markets, and poor soils and climate.

Superficially, one could judge the policy successful, if not cost-effective. Production targets have generally been fulfilled or overfulfilled, and land and human resources in disadvantaged regions are considerably greater than they would have been under a free market regime. But through the years, critics have pointed out the fallacy of using peacetime production to measure wartime preparedness. Under a protracted blockade, major dietary adjustment would be necessary (i.e., fewer animal and more vegetable products). Input combinations would have to be altered on a massive scale. In particular, farming methods would have to be modified dramatically if imports of fuel, fertilizer, pesticides, protein feed, and spare parts were cut off. On the other hand, production of a basic need such as food would probably receive an infusion of nonfarm labor and equipment to offset any shortage of more specialized inputs. Food security policy was not systematically grounded in such an analysis of economic vulnerability and flexibility, and it is difficult in retrospect to identify much genuine national security benefit from costly import restrictions and price supports (Bolin et al. 1986, Hedlund and Lundahl 1985, MOA 1989, Molander 1988).

It should be stressed that keeping arable land open is a form of insurance against risks and uncertainties, but maintaining high peacetime yields does not necessarily contribute to food security. A policy that promotes more land-extensive production methods, through the substitution of land for imported agrochemicals, is therefore likely to enhance security. The overall effect of agricultural policy before 1990 was the opposite.

A fundamental rethinking of national food security has been under way since the mid-1980s. In part, this has meant reorienting policy to crisis scenarios that are more plausible in the diplomatic-political-military environment of the late twentieth century. The notion of Sweden's holding out with no imports of food or farm inputs for several years has been officially abandoned, and new thinking centers on cost-effective ways to ensure country-wide food supplies under a blockade. The capacity to satisfy peacetime demand, which is strongly biased toward luxury foods, has given way to two other priorities: maintaining at least a minimum arable land base and stocks of milk cows in every region, and adding to stockpiles of nonperishable foods and critical farm inputs.

Government expenditure for stockpiles increased by 150 percent between 1981 and 1986, even before a general agricultural policy reform was on the agenda. A study done for the Swedish Defense Research Establishment in 1988 provoked a storm of controversy by concluding that a well-designed farmland/stockpile strategy could achieve national food security for about 5 percent of the cost of the existing agricultural policy (Molander 1988, Uhlin 1989).

We note that the greening of agricultural policy has a connection to food security, since recent measures to encourage reduced-input and organic farming (described in Chapter 7) limit Swedish dependence on imported biocides, feeds, and fertilizer feedstocks. That has not been an explicit objective of these measures, however, and their aggregate impact has been minor.

Table 4.1. Disposable income and wealth of Swedish households, by occupation, 1982 (in US$)

	Farmer[a]	Small business owner	Blue-collar worker	Low-level official	High-level official
Disposable income	13,800	15,200	17,200	20,200	25,800
Net worth	146,700	39,500	22,000	39,000	65,800

Source: NM 1989:78–79.
Note: Figures in Swedish kronor converted at the 1982 exchange rate of SEK 6.0 = US$1.
[a]Average for farms with 20–100 ha holdings.

The Income Goal: Equity for Family Farmers

Economic logic suggests that in a dynamic market economy—like Sweden's between 1945 and 1975—price incentives will stimulate a gradual flow of resources from sectors with low implicit wages and returns on assets, such as agriculture, to sectors with higher wages and returns, such as manufacturing. Given sufficient time, this process should go a long way toward equalizing incomes across sectors without any special government intervention. Swedish farm people have indeed been mobile, especially intergenerationally. The agricultural labor force contracted by 86 percent between 1950 and 1990, and there has been a high degree of long-term occupational mobility within farm families.

More than two-thirds of farm households' income now derives from nonagricultural sources, and although agricultural income is strongly correlated with farm size, overall farm household income is not (Figure 3.5). The complex class status and multiple income sources of contemporary Swedish farm households have raised questions about the operational meaning of equity and the concept of parity pricing. Table 4.1 compares the disposable income and net worth of farm households who own moderate-sized farms with other occupational groups. Much could be said about these figures, but two facts stand out in comparing the typical farm household with skilled blue-collar households, the reference group used in parity formulas. While farmers' income remained about 20 percent below that of workers, their wealth—primarily land assets—was more than six times greater. It is also striking that while the average farm household earns only about half as much as the family of a high-level official, its wealth is more than twice as great. This "live poor, die rich" phenomenon reveals the fundamental flaw in a policy that relies on artificial pricing to raise farm incomes: supports become capitalized in asset values. Ironically, this process was reversed after 1982, as depressed real farm prices led to negative real returns on farm assets.[12]

Money income is, of course, a weak indicator of equity and well-being. Self-employment has a strong attraction for farm people, even though it entails longer work hours and greater health and safety risks than most other occupations. In the past, farmers also benefited from certain tax breaks (e.g., lightly taxed tractor fuel can be used in the family car), lower living costs (e.g., free parking and low-cost heating with wood from the woodlot), and costless amenities such as beautiful surroundings and hunting opportunities.

There is no precise measure of the agricultural policy's impact on net farm incomes. The OECD producer subsidy equivalents presented in Chapter 2 show that price and income supports boosted short-term farm receipts by roughly 40 to 60 percent in recent years, and Erik Fahlbeck (1990:3) estimates the annual monetary transfer to the farm sector at $13,600 per farm, or 70 percent of agricultural value added. But these figures measure the effect on gross farm revenues, not on net farm income, which is the policy's equity norm.[13]

Logic and empirical evidence argue that import levies and regulated prices are extremely inefficient means of pursuing income equity, both among different types of farmers and between farmers and consumers. This is true, first, because price supports accrue to farms in proportion to their output. This favors producers with large tracts of fertile soils and favorable climate. Three-fourths of the revenue enhancement from Swedish price supports goes to the largest one-fourth of farms. Not surprisingly, they are heavily concentrated in the plains regions. A growing number of the big beneficiaries are quasi-capitalist or capitalist enterprises, not the genuine family farms that traditionally were the intended beneficiaries. Further, the system discriminates against producers of nonregulated commodities, such as fruits, vegetables, sheep, and horses. To cope with the defects of price supports, the state has had to take further measures—the Norrland support, low-income

[12] Estimates of the return on farm assets use the market wage for skilled workers to measure the implicit cost of self-employed farm labor.

[13] OECD's producer subsidy equivalent is an imprecise measure of the difference between current farm revenues and what they would be in a deregulated environment, since it holds farm output mix and input proportions constant. These would change substantially if price supports were removed.

supplement, and milk price premium—which cost taxpayers US$169 million in 1988/89 (SS 1991a).

Second, since expected future price supports are capitalized in land values, this is an inefficient way to transfer income from consumers. It benefits established producers who own their land outright. But higher land prices to tenants and entry-level farmers who seek to purchase land are a cost rather than a benefit. This phenomenon is doubly regressive, since wealthy landowners are subsidized at the expense of both consumers and land-poor farmers. This situation was stood on its head toward the end of the 1980s, when the land market was depressed by falling real farm prices and anticipation of price deregulation. Indeed, some large, efficiently managed farms were pushed into bankruptcy by the conjunction of high interest rates with declining cash flow and asset values. Farm bankruptcy has been rare in Sweden, and it was certainly not the state's preferred way to encourage the outflow of excess resources from farming (Drake 1989, Fahlbeck 1990).

Third, there is a widening gap between the prices paid by consumers and received by farmers. Figure 4.2b shows this trend clearly. By one estimate, each additional $100 consumers paid for food in 1985 raised farmers' net income by a mere $3, which is astonishingly cost-ineffective. Twenty-nine percent of the incremental cost went to farm inputs, 5 percent to imported food, 44 percent to food processing and distribution, and 19 percent to consumption taxes (MOA 1987a:44). Import protection and market power in first-stage food processing put agribusiness in a position to capture a large share of the gains each time farm prices are raised. After 1985, the requirement that producers pay 60 percent of export-dumping costs further eroded the link between consumers' expenditure and producers' income. Fahlbeck (1990) estimates that farm programs cost consumers and taxpayers 1.6 to 2.0 kronor for every krona of added farm income. Direct or "decoupled" income transfers would have been far more cost-effective.

Economic Rationalization and Efficiency Goals

The agricultural rationalization goal, first articulated in 1947, should be distinguished from the broader issue of farm policy's impact on national resource allocation. The preceding analysis has already suggested several ways in which import protection and price supports distort overall allocation. The goal of rationalization is best thought of as improved dynamic efficiency: more rapid farm productivity growth and a reduction of static resource misallocation over time. In Sweden, agricultural rationalization for the most part connotes structural, technological, and managerial innovations at the farm level. As we have seen, the major objectives have been to facilitate the flow of excess labor out of agriculture and to keep the policy's cost to consumers and taxpayers within politically tolerable bounds.

If the political commitment to import protection and price supports is taken as given, the rationalization measures described in section 4.1 were quite effective at achieving their macroeconomic objectives for several decades. Total factor productivity grew exponentially at nearly 2 percent annually (if we omit hidden environmental costs),[14] technical innovation proceeded continuously, and the farm workforce declined by nearly 90 percent. Meanwhile, real farm output increased and dietary quality improved, even as expenditure for food fell from 28 to 19 percent of private consumption. Of course, powerful economic and technological forces were pushing in the same direction as policy, so such public measures as research, extension, investment grants, and land consolidation assistance are perhaps best viewed as complements rather than prime movers.

On the downside, rationalization's effects conflicted with other policy objectives, particularly regional development and environmental protection goals (discussed below).

The Consumer Goal: Quality Food at Reasonable Prices

The consumer goal, also a policy priority in the 1920s and 1960s, was formally incorporated into agricultural policy in the 1970s, when the term "food policy" began to be used. The weight given consumer interests was sharply increased in the mid-1980s.

According to the OECD's 1988 study of Swedish agricultural policy, "there is no doubt that Swedish consumers have access to food of high quality" (1988a:14). Hygienic standards are rigorous and chemical residues are low by international standards. The reduced competition that results from import restrictions somewhat limits the variety of processed and manufactured food products available to Swedish consumers, at affordable prices, however (Uhlin 1989).

Obviously, consumers pay higher food prices than they would under a free trade regime. The magnitude of the differential depends on the

[14] Total factor productivity is an index of the value of output per unit of combined inputs, weighted by their proportional contributions to production.

gap between domestic and international prices and the level of retail food subsidies. The domestic-international price gap has fluctuated widely since World War II, becoming exceptionally large in the mid-1980s (Figure 4.2). For a number of years, retail subsidies buffered consumers, but for most of the postwar period they have borne 75 to 85 percent of the policy's total cost. Retail price inflation is compounded by import duties on first-stage processed foods and an oligopolistic distribution network.

The OECD's (1988a) calculations for 1980–87 indicate that retail food prices in Sweden were 46 percent above the average of other European members and 121 percent above the non-European members. In 1986, consumers paid 56 percent more than they would have in a market open to international competition; in monetary terms, this cost differential was $240 per capita, or more than 2% of personal income.

Swedish economists contend that the OECD's measure understates consumers' welfare loss, since uneven import levies distort the pattern of domestic prices. This in turn distorts consumption away from the welfare-maximizing market basket of foods. Swedes consume less food in general and considerably less meat and other livestock products than they would if relative prices reflected their preferences and price distortions were consistent across commodities.[15] By analyzing the demand response to price changes and the degree of price distortion for different commodities, Fahlbeck (1989) concluded that the monetary value of welfare losses in 1986 was $350 per capita, or one-third more than the OECD's estimate.

Inflated food prices have a regressive distributional effect, since low-income households spend a larger fraction of income on food than do high-income households.[16] Measures to achieve equitable incomes for farm households thus contradict a basic principle of distributional equity. This helps explain renewed emphasis on the consumer goal in the 1980s.

It is no mystery that import restrictions and price supports are an inefficient way to transfer income. The cost in consumer welfare is far greater than the benefit to farmers, both because consumption is

[15] The claim that differential import duties cause dietary distortions is borne out by evidence from the period of consumer subsidies, when there was both an absolute and a relative increase in consumption of livestock products (Drake 1989). Note that a green nutritional motive was used to justify the termination of meat subsidies: consumers would be healthier if they consumed less meat.

[16] Free school lunches and a subsidy on retail milk sales until 1990 slightly offset this regressive effect.

distorted and because much of the consumer's added expenditure fails to reach farmers as income. By the estimate mentioned earlier, the policy cost consumers and taxpayers 1.6 to 2.0 times what farmers' gained in income in the mid-1980s.[17] Ironically, it is consumers in food-importing countries who benefit from the dumping of Swedish surpluses on the world market.

Regional Development

Sweden has sought to maintain economically viable agriculture in all regions for both national security and broader socioeconomic reasons. In sparsely populated and geographically disadvantaged areas, this goal runs afoul of two inexorable trends. First, the public cost of keeping agricultural land rents above zero and returns to farm labor competitive has risen continuously. Second, shrinking farm activity erodes agriculture's capacity to invigorate the rural economy through interindustry linkages and local spending of farm income. The "critical mass" of farm production needed to sustain farm-related businesses has already disappeared from hundreds of communities. Countertrends include the growth of part-time off-farm employment, new niche opportunities in value-added commodities, and complementarities with the growing tourism and recreation sector (GBD 1984).

The general scheme of price supports, as indicated above, disproportionately benefits regions with a long growing season and large expanses of fertile land. In 1986/87 these payments averaged US$860 per hectare in the southernmost plains, versus only $420 per hectare in the north. The state has tried to counteract this bias with directed supports, including the milk price premium, Norrland income supplements, and numerous smaller initiatives directed at disadvantaged regions. These policies accounted for over half the agriculture ministry's budget in the 1980s and were increasingly important to the far north, where they totaled $301 per hectare in 1986/87 compared to a production value of only $452 at world market prices (NAMB 1988a, Uhlin 1989). Even together, however, these correctives do not offset the regional bias of price supports. In 1986/87, the plains region's share of aggregate agricultural supports was larger than its share of farm production (Table 4.2). Moreover, domestic supports cost more than the world market value of Swedish farm production.

[17] The monetary and welfare costs of import protection are very sensitive to domestic-international price relationships. Thus, the welfare cost grew sevenfold between 1980 and

Table 4.2. Regional agricultural production and support, 1986/87 (in millions of 1987 US$)

Region	Share of production[a]		Share of support: Price[b]		Share of support: Directed[c]		Total	
Southern and central plains	$903	60%	$1,188	67%	$9	9%	$1,197	64%
Southern and central forest regions	$453	30%	$459	25%	$18	19%	$477	25%
Norrland	$142	10%	$137	8%	$68	72%	$205	11%
SWEDEN	$1,498	100%	$1,784	100%	$95	100%	$1,879	100%

Source: NAMB 1988a.
Note: The 1986/87 exchange rate of SEK 7.3 = US$1 is used.
[a] Production valued at current world market prices.
[b] Values calculated using domestic prices minus world market prices.
[c] Principally the milk price premium, investment subsidies, Norrland support, and social supports.

The economically optimum production unit in forest and northern regions is smaller, more diversified, and more labor-intensive than in the prime grain-producing areas. Rationalization measures have further reinforced the competitive disadvantage of farms in these regions. State-funded research has disproportionately benefited larger-scale, more specialized, and more mechanized production systems. Some capital investment has been subsidized as part of rationalization policy, and the income tax code lets capital expenditure off lightly while placing a heavy burden on labor.[18] In sum, although the number of surviving farms and the extent of open land in the northern and forest regions are far greater than they would have been under a laissez-faire policy, the goal of economically sustainable agriculture in disadvantaged regions has not been pursued efficiently.

Environmental Protection and Nature Conservation

Environmental protection and nature conservation did not become official agricultural policy goals until 1984, when a host of green initiatives burst on the scene (described and evaluated in Chapters 6 and

1986 because domestic prices rose while international prices fell (Bolin et al. 1986; see also Figure 4.2).

[18] Capital investments have been eligible for investment tax credits and accelerated depreciation and interest write-offs, whereas payments to labor are subject to income and social security taxes plus heavy supplements to cover pension, health, and other benefits.

7). What should be underlined in this historical overview is that the need for environmental protection, nature conservation, and animal-rights measures has roots in policy failures, especially the effects of price supports and rationalization measures. Two among many possible illustrations are surface water eutrophication and the loss of biologically rich wetlands and meadows to rowcrop production. In the first case, pollutants from fertilizer and manure have been increased by extension recommendations supporting intensive production, by tax credits for investment in large-scale livestock confinement facilities, and by crop prices that artificially boost the payoff to fertilizer. In the second case, removal of grass-lined ditches, tree islands, and stone walls has been subsidized, and homogenization of landscapes is part and parcel of the simplified rotations made profitable by selective price supports.

4.3 On the Threshold of Policy Reform

Since its inception, Swedish agricultural policy has had multiple goals. At first glance, the policy and the farm sector seem to have performed fairly well over much of the postwar period:

- Farmers' incomes moved close to equality with nonfarm reference groups
- Most of the farm workforce successfully shifted to nonfarm occupations
- Peacetime self-sufficiency was attained in nearly all basic foods
- A production base was maintained in every region, and very little farmland was lost after the mid-1960s
- Labor productivity and land yields rose several fold
- Consumers had a safe, high-quality menu of foods
- Food expenditure declined as a fraction of consumers' income
- The policy's fiscal burden was quite small, compared, for instance, to that of the European Community or Japan.

The full story, as we have shown, is more complex and contradictory. Agricultural policy can take only partial credit for most of these accomplishments, and several of the positive results had downsides. The core instruments were neither target-efficient nor cost-effective and actually worked against some of the policy's goals.

Times change. For all their inefficiencies, import protection and price supports had a fairly low economic cost and were politically tolerable from the 1940s through the 1960s, so long as excess production was small and consumer income was growing rapidly. The same was true in the 1970s, when international food shortages narrowed the gap between domestic and world prices and raised the priority of food security. But these tools became blatantly counterproductive with the chronic surpluses and depressed international prices of the 1980s. Mounting evidence of the policy's adverse regional and environmental effects compounded the sense that more than minor tinkering was needed. A critique was advanced by prominent academics (see especially Bolin et al. 1986, Hedlund and Lundahl 1985) and began to be echoed in the mass media (Westberg 1988).

By 1984, the Social Democratic government was championing consumers' interests and arguing for a market-oriented deregulation of the food economy. By the time of the 1985 food policy legislation, this position had growing acceptance in the Liberal and Moderate parties. Although the 1985 food policy legislation itself was indecisive, it was preceded and followed by an outpouring of official studies that cast a critical eye on past methods of dealing with issues ranging from surplus grain disposal to income support, food security, rural economic decay, water pollution, and endangered wildlife habitat. It is telling that many of these investigations originated outside the agricultural establishment (GBD 1984, MOD 1988, Molander 1988, NEPF 1986). In 1986, the Uruguay Round injected the prospect of international trade liberalization into the Swedish farm policy debate.

Similar conditions and debates were widespread among the advanced capitalist nations in the latter half of the 1980s, but Sweden was first to take the leap to a new policy. In the fall of 1988, the government formed a food policy working group and challenged it to shape a new agricultural vision and a new policy.

5

The Negotiated Economy and Agricultural Policy

The contradictions inherent in Swedish agricultural policy became amplified in the 1980s. In this chapter we attempt to answer two questions about the policy regime: Why was an interventionist policy that gave privileged treatment and inordinate influence to farmers and agroindustry kept in place so long, despite ample evidence of its defects? What forces, inside and outside the food system, finally brought Sweden to the verge of a fundamental policy overhaul at the end of the 1980s? Our analysis probes three core governance processes: major agricultural legislation, farm price negotiations (and, by implication, the level of import protection), and the detailed design and administration of policy by state agencies.

5.1 Agricultural Policy in a Negotiated Economy

A fully developed theory of the capitalist state is beyond the scope of this book, but it is important to locate agricultural policy within an analytic framework encompassing the larger political economy. This is preferable to treating agricultural politics as an unopened "black box" or proceeding from simplistic assumptions, for example, that the visible hand of the state is rationally guided to optimize social welfare or alternatively that politics is little more than a contest among competing interest groups.

Alan Winter, in his description of the political economy of agricultural policy making, captures some of the complexity and ambiguity we hope to illuminate:

> The principal elements are a public opinion that abhors rapid change and exhibits a great sympathy for the virtues of country life, and a rate of economic progress that constantly threatens traditional agricultural methods. This conflict provides a seedbed for intervention, the nature of which is determined by the interactions of the farm lobbies, the bureaucrats and the politicians. However, the relationships among these groups do not generate a stable equilibrium that defines the nature and extent of agricultural policy. Rather, the parties are engaged in an endless and uncertain game (in the technical sense), each "winning" some rounds and "losing" others, but all anticipating the next round and all buffeted by the same shocks from outside. (1987:286)

We are struck by the accuracy of Winter's description of a policy process centering on ceaseless interest group contention but also incorporating two other forces: political culture (public "sympathies") and general economic trends and conditions (the "rate of economic progress" and "shocks from outside"). But before we explore the forces determining Swedish farm policy, the social formation within which such forces operate merits further attention.

One school of social scientists conceptualizes the institutional nexus of Scandinavian capitalism as the "negotiated economy." We find this term and the relationships it connotes useful. The negotiated economy contrasts with the familiar "mixed economy," in which three characteristic institutions—competitive markets, representative government, and administrative rules (bureaucracy)—interact to shape resource allocation and income distribution. In twentieth-century Scandinavia, a fourth core institution—state-guided interest group negotiations—has taken on major allocative and distributive functions. Of course, negotiated economic relationships occur to some degree in all advanced capitalist nations, but in Scandinavia "homogeneous populations, relative symmetry of power in the labor-capital relationship, and long history of compromising, integrating and mediating" have facilitated the evolution of a distinctive mode of capitalism (Nielsen and Pedersen 1990: 3). Esping-Andersen (1990) alludes to similar causal forces in his analysis of Scandinavian citizens' wide-ranging "inalienable rights" and the legitimation of an interventionist social democratic state.

In such an economy, according to Klaus Nielsen and Ove Pedersen, "the decisionmaking process is conducted via institutionalized negotiations between the relevant interested agents, who reach binding decisions typically based on discursive, political or moral imperatives rather than threats or economic incentives, even if such threats and rewards . . . might be essential elements of the framework around the negotiations" (1990:3). The four mechanisms—markets, bureaucratic rules and regulations, electoral politics, and interest group negotiations—can be thought of as mutually constraining one another and jointly determining economic outcomes.

The relatively autonomous Swedish state initiated negotiated relationships in the late nineteenth and early twentieth centuries, for instance between trade union and employer federations and between tenant and landlord organizations. In some cases, central or local government agencies have been directly involved in the bargaining process, an arrangement commonly termed "democratic corporatism." This arrangement places the state in the contradictory roles of representing the public interest and also mediating among narrow private interests. Corporatist arrangements in Sweden actually predate modern parliamentary democracy, and some of the most important corporatist relationships were essentially inventions of the state administration.[1] In agriculture, for example, the bureaucracy prodded encompassing interest organizations into existence in order to facilitate national price negotiations. Micheletti (1990) describes this as "sponsored pluralism," stressing that most citizens accept de jure the direct involvement of interest groups in policy decisions and public administration (see also Rothstein 1988).

A central proposition of democratic corporatist theory is that neither the state nor private interest organizations are fully independent actors: "The state delegates powers and participates in decisionmaking processes without full authority," while the interest organizations are "integrated in the political process in a stable, long-term manner" through a nexus of moral, cultural, and discursive ties that temper self-interested behavior (Nielsen and Pedersen 1990:7, 13). In the terms of Albert Hirschman's *Exit, Voice, Loyalty* (1971), the official emphasis on "voice" expressed through interest organizations, and on "loyalty" developed through stable, long-term, negotiated relationships, has been reinforced by the practical impossibility of "exiting" from corporatist relationships.

Narrow self-interests and group interests are "tamed" but not purged by participation in the negotiated economy. Indeed, it has often been observed that interest organizations employ Trojan horse tactics, rationalizing self-interested positions with rhetoric that upholds a larger collective purpose. An intriguing implication of such hypocriti-

[1] Democratic corporatist arrangements, found in many parliamentary democracies, should not be confused with authoritarian corporatist regimes, such as those of Nazi Germany and Fascist Spain.

cal public posturing is described by Jon Elster: "Strategic use of impartial arguments [by interest groups] tends toward more 'righteous' decisions. If consideration to voters or others makes it hard to present demands based on pure self-interest, the public interest has improved prospects. The [groups'] optimal impartial arguments will often reflect self-interest in a diluted form that considers other groups. This may be considered an example of the *civilizing power of hypocrisy*" (1991:9; emphasis in original).

The negotiated economy is continuously evolving, as the four core institutional arrangements "compete, disturb and eventually supersede each other" (Nielsen and Pedersen 1990:3). We believe national economic negotiations have indeed been "disturbed" and are losing their hegemonic role—most obviously to market forces, but also to new administrative procedures, and, in this period of fluid political party loyalties, to electoral politics and to public opinion, expressed in many ways.[2]

The neoclassical economists' principal tool for political analysis, public choice theory, interprets the historical evidence quite differently from the negotiated economy school. The neoclassical position is that political actors (with rare exceptions) are rational and optimize their self-interests. This assumption extends to politicians and bureaucrats as well as to private "principals" and their "agents," the interest organizations. An economic interest group, such as dairy farmers, mobilizes and employs political resources (lobbyists, election contributions, media advertisements) to gain and retain income streams ("economic rents") through favorable state intervention. An extreme version of public choice theory posits the existence of a "political marketplace" wherein economic and other interest organizations demand and compete for favorable policy treatment. Self-interested politicians and bureaucrats are the responsive suppliers (Anderson and Hayami 1986, Olson 1965). This is one perspective on the "endless and uncertain [political] game" referred to by Winter above.

A public choice interpretation of Swedish agricultural policy holds that the political marketplace is monopolized by a collusive "iron triangle" made up of farm organizations, the agricultural bureaucracy,

[2] Public opinion, especially as expressed in opinion surveys, is a social construct subject to manipulation and not an expression of autonomous individual sentiment. Expressed opinions, furthermore, are merely the outward, quantifiable sign of deeper, interconnected forces. Obviously, one force is the survey respondent's perceived self-interest, which public choice theorists stress. A more subtle force is moral convictions, such as solidarity and altruism, which pass through the public choice filter. Swedish public opinion is considered further in section 5.6 and in Chapter 6.

and the Riksdag's farmer-dominated Permanent Agriculture Committee (see Bolin et al. 1986, Rabinowicz et al. 1986). As we shall elaborate, some elements in the public choice story are helpful in understanding Swedish agricultural policy. But in its extreme version, public choice seems to us a misleading caricature of Sweden's complex and changing political reality.

In this communications age, how citizens receive, interpret, and act on "news" is a crucial but exceptionally complex subject. The Swedish print and electronic media transmit and also shape political discourse, thereby influencing policy outcomes. Indeed, the private and state-owned media and influential people within their hierarchical organizations should be viewed as political actors in their own right. In Sweden, the media's agency role derives in part from the close association between political parties and most of the major daily newspapers; in addition, the LRF owns four wide-circulation periodicals.[3]

The mass media are a focus of intense competition among political actors seeking to promote their agendas and discredit opposing viewpoints. Television in particular affects every political issue with its capacity to dramatize certain issues relative to others, to enhance the image of some actors relative to others, and to influence everyone's tactics. Political actors today have no choice but to commit resources to the media game of influencing and responding to public opinion (Petersson and Carlberg 1990).

5.2 The Grand Policy Design: Private Interests, Collective Commitments

The formation of Sweden's interventionist agricultural regime during the 1930s economic depression can be interpreted on the one hand as economic crisis management and on the other as a political compromise that enabled a "red–green" (Social Democrat–Agrarian) coalition to take power from the bourgeois political parties. Swedish experience confirms Petit's (1985) view that general economic and technical conditions and the exigencies of political compromise and coalition building are basic boundary conditions within which agricultural policy is shaped.

[3] An important feature of Swedish political culture is that a large majority of Swedish adults read daily newspapers. The proportion is more than double that of the United States, suggesting somewhat less dependence on passive television viewing for current events information.

In keeping with the historic compromise, farmers and their supporters were, and still remain, well represented on the Riksdag's Permanent Agriculture Committee. The proportion of committee members who listed farming as their main occupation was 75 percent in 1940, 85 percent in 1960, 33 percent in 1980, and 41 percent in 1988 (Micheletti 1990:111).

Farmers' representatives also made up half the commission that proposed Sweden's first comprehensive agricultural policy, with its income parity goal, in 1947. As mentioned, Swedish citizens sanction overt interest group participation in policy design. Farmers' representatives on subsequent commissions ranged from a high of 44 percent in the 1970s, when the Center (formerly Agrarian) party reached peak influence, to a low of zero in 1988–89, when the reform-minded Social Democratic government created a working group consisting only of Riksdag members and civil servants (Andersson 1987).

Despite the breakup of the red-green coalition in 1957 and a continuous decline in the farm population and in agriculture's economic importance, agroindustrial interests were able to maintain a strong political presence favoring the status quo. This can be interpreted as the result of two principal factors. First, the Swedish Farmers' Federation (LRF), formed by a merger between the Swedish Farmers' Union and the agricultural cooperative associations in 1971, became, with state encouragement, the farmers' encompassing interest organization. This concentration of organizational capital helped counteract farmers' waning electoral influence by improving their ability to resolve factional conflicts and speak publicly with one voice. Second, economic growth in the input supply, food processing, and distribution industries offset falling farm numbers. While agriculture's share of the Swedish workforce is now less than 3 percent, the agroindustrial complex as a whole still employs about 11 percent. The economic and political bonds between farmers and agribusiness are strong. LRF's member cooperatives own most capital in input supply and primary food processing; moreover, secure markets and profit margins for agribusiness have been dependent on agriculture's continued protection from foreign competition. Finally, most agribusiness wage earners belong to trade unions that are linked to the Social Democratic party. Some believe this increased the SAP's tolerance of agricultural protectionism. (Andersson 1987, Bolin et al. 1986, Micheletti 1990).[4]

[4] As we have noted, class analysis is not the most useful point of entry for agricultural policy analysis. The common interest of agribusiness wage earners' and self-employed farm-

Farmers and agribusinesses are certainly rent seekers who engage in politics to protect the income streams and asset values secured by domestic price supports, import restrictions, and a host of other measures. These rents are essentially transfers from Swedish consumers and taxpayers at large. The public choice school interprets the durability of these measures, in vastly changed conditions from the 1930s, as a textbook case of Mancur Olson's "theory of collective action." Olson (1965) postulated that numerically small and cohesive economic interests, with much at stake, have a superior capacity to maintain effective political organization and are able to dominate large, heterogeneous groups such as consumers and taxpayers. These two groups are notoriously difficult to organize: their individual stakes in particular policy measures are small, and there is a strong tendency to "free ride" on the political effort of others.[5] These analysts also point to the LRF's public relations campaigns, which secured broad popular support for the status quo by successfully portraying farmers as virtuous, hardworking protectors of Sweden's food supply and its *levande landsbygd* (living countryside) (Bolin et al. 1986, Rabinowicz et al. 1986, Uhlin 1989).

The public choice story of interest group politics has considerable persuasive power. It caricatures the Swedish legislative process, however, by neglecting or misconstruing the way inherited institutional arrangements, public discourse, and ethical commitments affect political actors' behavior.

In Sweden's pluralistic interest group competition, agroindustrial interests have not monopolized legislative debates or outcomes. The agricultural commissions, as expected in a negotiated economy, include countervailing voices: political parties not beholden to farmer constituents, and consumer and labor representatives.[6] On the 1960s commission, for example, the majority opposed income parity and it was dropped as the major policy goal. Legislation is submitted to the Riksdag by the governing party (or coalition), which in the case of the

ers' in maintaining protectionist farm policy exemplifies the interweaving of class interests that complicates a class analysis of the state.

[5] This interpretation may break down in a multiparty political system where some parties compete by claiming to speak for consumers and taxpayers. In the Swedish context, it was the SAP government that created a consumer delegation to participate in farm price negotiations and that more recently stressed consumer interests in its push for agricultural policy reform.

[6] The proportion of consumer and labor representatives on agricultural policy commissions has been: 1940s, 18 percent; 1960s, 29 percent; 1970s, 11 percent. In 1988–89, as mentioned, both producer and consumer representatives were excluded (Andersson 1987).

Social Democrats has not been an advocate, still less a rubber stamp, for farmers. The parliamentary debate is organized around the government's bill, informed by *remissvar* (written responses to proposed legislation), solicited from organizations representing a wide range of values and interests. Taken together, these procedures are designed to balance competing interests and encourage formation of a broad consensus behind legislation (not that it is always achieved). The governing party or coalition seldom attempts to force an unpalatable policy down the throats of the opposition. But consensus presupposes compromise on all sides, including the LRF and its backers in the Center party. This has been a key to the policy regime's legitimation and persistence over such a long time (Weaver 1987).

Citizens' capacity to influence farm policy is problematic. The legislative process is widely reported by the mass media, which both inform the public and feed public opinion back to actors directly involved in the policy competition. This does not imply that Svensson takes the time to follow farm policy debates in detail, much less that he is well informed about the hundreds of agricultural regulations and programs. Further, as Olson argued, poorly organized consumers and taxpayers suffer more seriously from principal-agent problems than do narrow interest groups.[7]

In contrast, LRF is a full-time agent for farmers, aggregating and promoting their interests. It has a large budget and many tools for political action: widely circulating journals, newspaper advertisements, television interviews, and public debates. When a dramatic protest is tactically opportune, LRF can also mobilize hundreds of tractors to occupy the streets of Stockholm. In contrast, the labor unions and political parties that claim to speak for consumers and taxpayers have many other priorities besides agricultural policy. Without consulting their constituents, these agents take positions on an array of policy issues that may affect agriculture, such as national security, regional development, income distribution, and environmental protection.

This is not to say that consumers and taxpayers alone suffer from agency problems, since representation has also become a thorny problem for farmers and agroindustrial interests. In recent years, the LRF has faced a defection of some members, and there are internal splits along a series of fault lines: economically advantaged versus disadvan-

[7]The principal-agent problem refers to the slippage that occurs when individuals or interest groups ("principals") cannot represent their own interests effectively and so must rely on agents such as political parties and lobbyists.

taged farming regions, full-time versus part-time farmers, conventional versus organic farmers, and so on. As mentioned, agribusinesses owned by producer cooperatives have also gradually increased their autonomy from the cooperatives. And in Sweden's multiparty politics, the Center and Moderate parties now split the farm vote (Micheletti 1987, 1990).

The SAP tolerated farm policies for decades, despite their negative economic efficiency effects and their burden on wage earners. We believe this tolerance was not solely a result of constraints imposed by political compromise and coalition building. Politicians, like their constituents, are subject to tensions among conflicting interests, values, and ethical convictions. Indeed, public choice theorists in their less doctrinaire moments acknowledge that political culture molds politicians, too. In early industrial Sweden, the democratic socialists rejected the Marxian stereotype of "the idiocy of rural life" and the Leninist disdain for the peasantry. On the contrary, most Swedish political leaders and urban dwellers, across party and class lines, had a place in the heart for country life and for the yeoman farmers they viewed as its stewards. These sentiments were bound up with a characteristic Swedish love of "Nature" that was cultivated by turn-of-the-century bourgeois publicists and has been documented by anthropologists Jonas Frykman and Orvar Löfgren:

> [Nature] came to stand for the authentic and unaffected in contrast to the artificial, manmade and commercialized milieux of the urban world. . . . Nature was animated into a symbol for Swedishness and a national fellowship above class boundaries . . . through [the publicists'] work they helped construct the myth of a traditional and national peasant culture. . . . The cult of a peaceful countryside and the old peasant life can be seen as an attempt to create a common national identity in a period of sharpened class conflicts in Sweden. . . . Both liberals and socialists used the image of an egalitarian peasant society as a model for the future. (1979:57–62)

At the time of the 1930s "cow trade," this residue of preindustrial values and solidarity with hardworking country cousins gave moral weight to the farmers' claim to economic support.

Positive feelings toward farmers and the countryside, though arguably diminished, persist today in conditions very different from those of the 1930s. The sentiment mingles self-interest with altruism (solidarity) and is sustained by a mix of childhood socialization,[8] personal experience, and media reinforcement. The element of self-interest is obvious. Food self-reliance is still considered a public good: insurance in case of an international food-supply crisis. But agriculture's perceived collective benefits go far beyond the food supply. Hundreds of thousands of Swedes commute to work daily through agrarian landscapes that give pleasure and relieve stress. Opinion surveys reveal that protection of an open and varied landscape is now the voters' number-one motive for supporting Swedish agriculture. In a strongly leisure-oriented culture, millions return each year to bucolic rural villages to relive *den gamla goda tiden* (the good old times), dancing around the maypole, picking flowers in the meadow, visiting the open air museum, and shopping at the farmers' market. Vacation at *den lilla stuga på landet* (the little cottage in the countryside) is a basic Swedish image of the good life. It is fitting that when Finance Minister Kjell-Olof Feldt, the "gray eminence" behind agricultural policy reform, resigned under pressure in February 1990, he joked that it was time to fix up his cottage. Satisfying all these nonfood wants, of course, requires public support to sustain a viable agriculture (AERI 1992, Micheletti 1990, Westberg 1988).

Popular preferences regarding food security, aesthetics, and recreational amenities have not only affected policy goals but have also created a bias in favor of particular means, especially supports targeted to family farms and sparsely populated regions. Neoliberals view this instrumental bias as irrational, and from a narrow economistic perspective it is. Yet the bias is also evidence that the public interest has complex and contradictory meanings and is colored by emotions. This is consistent with Etzioni's observation that "people typically choose means largely on the basis of emotions and value judgments and only secondarily on the basis of logical-empirical considerations" (1988:xi).

Etzioni's I–We paradigm, mentioned in Chapter 1, captures the complex web of self-interest and community identity, as well as the "balance between pleasure and moral conduct" that typifies behavior in a "highly integrated community" such as Sweden's (1988:85). The 1930s worker-peasant alliance was grounded in objective conditions: wage earners and yeoman farmers shared low material living standards; in late-industrializing Sweden, most workers still had one foot

[8] Hardly a Swedish child grows up without being exposed to the romance of the countryside, as portrayed, for example, in Astrid Lindgren's tales of Bullerbyn.

in the village.[9] Most urban Swedes embraced the populist metaphors of farmers as the *landets ryggrad* (nation's backbone) and the nation as the *folkhemmet* (people's home). These images fostered an inclusive social vision rather than class hostility (Micheletti 1990).

The 1947 commitment to farm income parity was consistent with the wage solidarity principle introduced into centralized wage bargaining at roughly the same time. And just as popular sentiment favored higher wages—not welfare payments—for workers in low-productivity occupations, so too it favored farm price supports, rather than more efficient transfer payments, to reach farm income parity. The wage-earning majority empathized with farmers' desire to earn their keep and not be treated as *postgirobönder* (postal check peasants). Notwithstanding Sweden's international reputation as a generous welfare state, a guiding principle of its social contract has been to encourage self-help rather than long-term economic or psychological dependency (Micheletti 1990).[10]

Our emphasis on the ideological foundation of public support for agricultural policy should not obscure crucial material factors. During the thirty-year economic expansion that followed World War II, steadily rising nonfarm wages more than offset the costs of agricultural policy. Diet improved and food took a smaller bite out of household income.[11] There was also an enormous increase in leisure time, part of which was used to enjoy the amenities of the agrarian countryside. If there is a political puzzle, it is that popular support for farmers and agriculture did not erode quickly in the 1970s when nonfarm income stagnated, food inflation escalated, and an increasingly industrialized agriculture began to lose the qualities that had attracted public sympathy (Rabinowicz 1991a, Westberg 1988). One explanation of the inertia in public sentiments is simply that citizens were poorly informed about the extent to which agriculture was being industrialized and the

[9] Worker-peasant solidarity in the 1930s was based on the earlier socioeconomic structure of Swedish villages, which had pitted both peasants and rural wage earners against wealthy landowners.

[10] The self-help principle has roots in Swedish Protestantism and conceivably even in the Viking heritage. The principle was embodied in many Social Democratic policies and quintessentially in the active labor market policy. More than any other industrial nation, Sweden has channeled public assistance for unemployed workers into active measures, including employment counseling, retraining, and relocation assistance. Along similar lines, Rothstein (1992) and many others contend that a reason for the welfare state crisis since the mid-1970s has been growing entitlement dependency and a weakening of mechanisms encouraging individual self-reliance.

[11] By dietary improvement we mean increased variety and affordability of foods, especially outside the growing season. We pass no judgment on the nutritional quality of diets.

role of farm policy in the process. But the lag between changes in objective conditions and in public opinion also supports Etzioni's thesis that "moral behavior is 'stickier' than a-moral behavior" (1988:68).

We have argued that such agents as politicians, bureaucrats, and lobbyists are not separate from society: they are socialized in the same cultural and institutional milieu as their constituents. Like Finance Minister Feldt, they have their cottages in the countryside. Thus, at least in the Swedish case, it rings false to claim that cynical politicians manipulate public opinion without being reciprocally affected by it. Winter describes a mix of expediency and commitment which helps explain the inertia behind an increasingly ineffective agricultural policy: "Governments seek to prevent, and certainly never initiate, events that significantly cut the welfare of any particular group in society. As well as meeting the community's sense of fairness, [such a] 'conservative social welfare function' is also attractive to policy-makers in that it probably promotes social harmony and political quiescence" (1987:292).[12] Thus the governing Social Democratic Party, though it had wanted to reform agricultural policy for several decades, held back from forcing through changes that lacked broad support from the public and the opposition parties.

5.3 The Rationalizing Impulse

It cannot be said that Swedish agricultural policy is founded on a clearly articulated public interest, much less a fully specified social welfare function. Nevertheless, there has been an impetus toward instrumental rationality, and this has been a countervailing force—sometimes weak, sometimes strong—against both rent-seeking interest groups and subjective bias in the choice of policy instruments.

During the Social Democrats' long period in power, a self-conscious project was the modernization and rational organization of society, with guidance from an enlightened state. Ron Eyerman (1985) traces the SAP's social engineering orientation back to a core group of "rationalizing intellectuals" who advised politicians or were themselves

[12] In a personal communication (1991) Micheletti points out that the principle of the conservative welfare function was violated in Sweden a generation ago, when Social Democratic redistribution measures significantly penalized the upper classes. Non-Swedish exceptions to Winter's dictum can also be cited, such as the Reagan administration's social and economic policies that reduced both the relative and absolute living standards of low-income Americans.

politicians during the interwar years. Skepticism about free market capitalism ran high, as did optimism about the potential of well-crafted state intervention to solve "the economic problem" and promote social harmony. For these collective ends to be accomplished, the power of both markets and narrowly based interest groups had to be held in check. The institutions of the negotiated economy were the principal means (Esping-Andersen 1985, Rothstein 1988).

The role of the rationalizing intellectuals can be understood at two levels. As ideologists, they advocated progress: a vision of technologically modern, rationally organized, materially affluent society. This technorationalist vision was tempered by populist Social Democratic commitments to *folkhemmet* and to participatory institutions. Social Democratic intellectuals such as Myrdal represented one pole in the party's internal tension between technocratic and populist cultures. As practitioners, the intellectuals designed a complex framework of public institutions and policies to mediate class conflicts and promote long-term technological, economic, and social modernization. Thomas Anton described this aspect of Swedish statecraft as "an elite culture in which a highly pragmatic intellectual style, oriented toward the discovery of workable solutions to specific problems, structures a consensual approach to policy making" (1969:99).

The policy recommendations and supporting social theory advanced by Tage Erlander, Gunnar and Alva Myrdal, Ernst Wigforss, and their cohort in the 1930s and 1940s could certainly be interpreted as a rationalization of their own interests as technocratic empire builders. Eyerman, however, views these leading figures as largely altruistic, motivated by an "ethic that lies somewhere between responsibility and conviction" (1985:779).

Per Thullberg's (1983) analysis of postwar agricultural commissions and farm legislation also accords a central place to experts from academia and the nonagricultural civil service. Thullberg views them as relatively autonomous actors who opposed the agricultural establishment's parochial and self-interested orientation with their perception of society's interest. Although their influence rose and fell over time, and they never had the power to dictate policy, the experts did "structure the problem of agricultural policy," keeping economic efficiency on the agenda and supplying supporting analysis. Thullberg, like Eyerman, accepts that these "engaged intellectuals" were committed to the public good; but their recommendations reflected a particular teleology: a "programmatic striving for efficiency in production, . . . rational organization of production . . . [and] technology-tinged notions of rationality" (1983:1). This goal hierarchy gives little weight to such old-fashioned values as family farming and preservation of cultural-historic landscapes. It also implies a policy regime with considerable authority delegated to technical experts.

The notion of autonomous intellectuals is problematic, since most of the experts brought into the legislative process by Social Democratic politicians had party connections. This was true of both the academic economists who staffed the 1960s agriculture commission and the Finance Ministry officials who directed the 1989 food policy working group. However "independent minded" the experts might be, in the end their recommendations had to run the gauntlet of a multistage legislative process before becoming policy.

In view of these qualifications, it seems to us that rationalizing intellectuals have been a significant counterforce to agroindustrial interests and to peasant romanticism. Their influence was apparent in the 1947, 1967, and 1990 farm legislation, but it was minimal in 1977 when the nonsocialist coalition government kept independent experts off its agricultural commission. In the half century following Myrdal's 1938 critique of agricultural protectionism and price supports, Sweden's rationalizing intellectuals were not able to undo the basic elements of agricultural policy. Their lesser roles were to challenge the notion that overproduction improves national security, to stress consumer well-being as a prime indicator of successful food policy, and to promote rationalization measures that raise productivity and accelerate the flow of labor and and capital out of agriculture (Thullberg 1983).

In sum, intellectuals outside the agricultural establishment never dominated the policy agenda, but they helped to impose what Petit (1985) calls a "minimum degree of rationality" and Winter (1987) calls "a boundary of 'reasonableness' beyond which governments cannot stray." Ironically, agricultural rationalization was successful enough that a policy the intellectuals opposed was able to survive without collapsing under its own weight.

5.4 Corporatist Pricing Arrangements: Capture and Co-optation?

Price setting is a core process in a regime that centers on price supports and domestic market regulations. Sweden's three-party negotiations involve the National Agricultural Marketing Board, the Producer Delegation led by LRF, and the Consumer Delegation (CD).

Policy analysts disagree fundamentally about the answers to two agricultural pricing questions: Behind the façade of negotiations, where does real power lie? Whose material interests are served and at whose expense? To the public choice school, the answers are clear. Agroindustrial interests have been far better organized than consumers and have had bargaining leverage. The Marketing Board, staffed by bureaucrats with close personal and professional ties to farmers, was therefore naturally biased in their favor. Indeed, price setting is viewed as the power focus of the iron triangle (farm and agroindustrial interests, the agricultural bureaucracy, and the Permanent Agriculture Committee of the parliament). Although the committee was not directly involved in price negotiations, it shepherded price policies through the legislative process.

The Marketing Board's decisions are viewed as quite independent from the government, even though the cabinet designs the budget that has to cover the policy's fiscal cost and the Riksdag formally approves or disapproves both the budget and the negotiated price package. From the public choice perspective, the Consumer Delegation (not formed until 1963) was a rubber stamp whose typical acceptance of ever higher farm prices served to legitimate them. In sum, corporatist price setting shifted power from elected representatives to an alliance of economic and bureaucratic interests. The winners were producers of price-regulated commodities and protected food processing industries. The losers were consumers and taxpayers.[13] The support for this thesis comes mainly from price data. In the 1980s, retail food prices were twice what they would have been in a free market environment, and they increased much faster than the general cost of living (Bolin et al. 1986, Rabinowicz et al. 1986, Uhlin 1989).

We believe that the iron triangle metaphor oversimplifies and distorts reality in four ways. First, price negotiations always took place within limits set by the Riksdag. Between 1977 and 1985, for example, the formula connecting farm production costs and prices was specified in such detail that there was relatively little to negotiate about. At other times the negotiators had more latitude, specifically when the Riksdag wanted to weaken the LRF's influence by dropping the income parity formula, as in 1967 and 1985. Indeed, the 1985 legislation not only removed parity pricing on regulated commodities, but also nullified nominal price increases by forcing grain and pork producers to

[13] Taxpayers shared the burden of high farm prices when they caused excess production and required export subsidies and when subsidies were used to offset retail food price inflation.

absorb 60 percent of the cost of export dumping. After the Riksdag phased out most retail food subsidies as a fiscal austerity measure, the Consumer Delegation began to resist farm price increases more vigorously (Micheletti 1990, Rabinowicz 1991a).

Second, the version of public choice theory which asserts that encompassing interest organizations exhibit unmitigated self-interest in negotiations is disputable. The contrasting view, presented earlier, is that stable, long-term relationships with consumer and state representatives encourage interest organizations' "good citizenship" and tempers their one-sided pursuit of members' economic interests. Based on a historical study of the LRF, Micheletti concludes that "the fusion of interest representation and policy implementation has . . . made the interest organizations more flexible to changes and ready to assume a problem-solving role that goes beyond the representation of members. Frequently the interest organizations have accepted policy solutions which cannot be said to be beneficial to the members or the organization in the short run" (1990:6–7). Similar conclusions about the "taming" effect of corporatist arrangements on interest group behavior can be drawn from studies of other Swedish social and economic arenas and other north European countries (Elster 1991, Etzioni 1988, Hirschman 1971, Nielsen and Pedersen 1990, Olson 1982, Pestoff 1990, Rothstein 1988).

Third, some facts from the 1980s cast doubt on the image of LRF as a sturdy leg of the iron triangle (section 5.2 raised questions about the reliability of the bureaucratic leg). LRF's ability to aggregate and optimize members' economic interests via price setting was weakened by internal conflicts among groups of members and between principals (members) and agents (the organizational hierarchy). A few examples illustrate. Higher prices to feed grain growers meant higher costs to livestock producers, costs that could no longer be passed fully to consumers. High grain prices also stimulated output, which increased the cost of export subsidies and almost certainly led to spending cuts for other farm programs. Finally, producers of nonregulated commodities, such as vegetables and sheep, did not share the grain and oilseed growers' interest in import protection. Other examples abound (Micheletti 1990).

Polarized debates at LRF's recent conventions and defection of former members to insurgent organizations reflect the LRF's increasing difficulty in reaching a strategic consensus on pricing as well as on other economic and ideological issues. Micheletti makes the case that the LRF is embroiled in an organizational identity crisis that goes far beyond rent-sharing disputes over farm prices. One indicator is the growth of a green LRF faction that has joined environmentalists to oppose chemical-intensive farming and promote alternative agriculture (Hasund 1990b; Micheletti 1987, 1990).

Fourth, a more thorough accounting must be made of the Consumer Delegation (CD). Did it routinely ratify inflationary decisions reached collusively by the LRF and Marketing Board bureaucrats, as public choice theorists suggest? Was it preempted by legislation? Or was it an active player, successfully restraining price increases and promoting other interests of Swedish citizens?

Organizing to resist food price inflation has precedents in Sweden: as early as the 1950s, the trade union federations and the consumer cooperative union joined forces to investigate and speak out on high food prices. At that time, the majority of Swedes belonged to consumer cooperatives. The Center party's and LRF's opposition to the creation of a consumer delegation in 1963 reflects their expectation that it would have real bargaining leverage.

Micheletti emphasizes that, once created, the delegation only gradually took on a distinct identity and an activist role, provoked by persistent price inflation and growing concerns about food safety. In the early 1970s, a relatively passive CD was prodded by grass-roots protests over food prices, organized by the "Skärholmen wives" of suburban Stockholm. Once the income parity formula was removed in 1985, the CD had greater freedom to resist big price increases. Its representatives several times broke off negotiations over unresolved conflicts with the LRF.

The CD's delegates increasingly recognized that leverage at the bargaining table required supportive public opinion, so they put greater emphasis on public relations. The LRF, as noted, already had a sophisticated and well-financed public relations strategy. In the latter 1980s, the Social Democratic party and the major trade union federations joined the crusade against high food prices. In their tactics, farmers were never portrayed as the enemy of wage-earning consumers; right up to the 1990 food policy reform, which dismantled corporatist pricing, the CD accepted that some price increase was necessary to support farm incomes and ensure the provision of public goods such as food security and a "living countryside" (Micheletti 1990, Nyberg 1989).

Victor Pestoff (1990) views the Consumer Delegation as a successful state-sponsored institutional innovation that gave consumers a voice and an influence they would not otherwise have had. He stresses the trade unions' role in strengthening the delegation's resolve and its popular backing. Indeed, during the 1989–90 food policy debate, the chair and spokesperson of CD was an official of the Central Confederation of Professional Employees. Micheletti's comparison of Sweden with countries where farm prices are negotiated without formal consumer representation concludes that "the presence of a countervailing power in the Swedish negotiations is an important reason for the moderation shown by the LRF" (1990:179).

Skeptics contend that the CD was nearly impotent and that the LRF did not moderate its price demands to any significant degree, at least until the end of the 1980s when it feared the dismantling of farm policy. Rabinowicz (1991a) sounds a familiar refrain: "well organized producers" have a systematic advantage over "badly organized consumers." She is correct to stress that the CD had no organic tie to consumers and that the consumer cooperative movement did little to strengthen the connection. The CD's public relations budget could not match the LRF's. And the anomaly of food industry representatives on a consumer delegation created conflicts of interest. The skeptics note that precisely when the CD supposedly became a force for price containment in the mid-1980s, Swedish food price inflation accelerated (Rabinowicz 1991a, Rabinowicz et al. 1986).

In our view, the Consumer Delegation had little restraining influence on farm prices for most of its existence. The strength of its resolve and influence was closely related to the level of retail food inflation and the negotiating flexibility allowed by legislation. The CD, however, became a tougher bargaining agent in the latter 1980s, reinforcing its "voice" with a will to "exit" from negotiations. In particular, delegation spokespeople learned how to represent consumers' interests in the media and in the food policy debate. In 1989, they called for a demand-guided farm policy that would eliminate corporatist pricing—and their own raison d'être. During these years, the CD also became a green force, advocating stronger protection of food safety, restriction of agrochemicals, and action to preserve farming in sparsely populated regions (Nyberg 1989). In sum, the record suggests that negotiated economy theorists overstate the Consumer Delegation's impact on price setting, but that in the critical years after 1985 the public choice theorists underestimate its influence on pricing and a range of green policy issues.

5.5 Ground-level Policy Implementation: Legitimated Collusion

Implementing agricultural policy entails interpreting and enforcing thousands of detailed regulations and providing a multitude of services to farmers. Before 1990, these functions were carried out in Sweden by two principal national agencies, numerous operating divisions, and a host of local subagencies and participatory boards. The case that agricultural policy is a collusive arrangement between farm interests and the bureaucracy is strongest at this ground level. But in Sweden this is not a devious relationship: a core feature of the negotiated economy is the direct involvement of interest organizations. Especially since farmers' well-being has been a central policy goal, it would be rational for their agents to participate in instrument design and implementation at this level.

As in several other European nations, professional civil servants in Sweden have been widely respected for their integrity and expertise. Administrative agencies are given considerable latitude in policy execution, and the minister of agriculture has little legal authority to intervene in many aspects of administration. Winter believes that governing political parties have a vested interest in depoliticizing administrative decisions and so turn them over to "the experts": "Governments deliberately establish bureaucratic-cum-technical procedures . . . because . . . by appealing to experts and couching the discussion in highly technical and apparently objective terms, they can take a lot of the heat out of an issue" (1987:297).

Sweden is typical in that many agricultural bureaucrats come from farm backgrounds and train alongside future farmers in specialized secondary schools and at the University of Agricultural Sciences. Shared backgrounds and jobs that involve day-to-day collaboration with farmers and farm organizations foster a work ethic whereby doing the job well means doing well by farmers. Farm organizations are indeed enmeshed in the design and execution of policy: "The state relies on the expertise of the interest organizations for its street-level agricultural bureaucracies. It needs the support of individual farmers, the policy *takers,* for successful implementation of policy" (Micheletti 1987: 169; emphasis in original).

An empathetic bureaucracy and the farm organizations' influence in street-level policy implementation are consistent with the iron triangle interpretation. Olson (1982) predicts a long-run economic decline for nations that succumb to the "institutional sclerosis" of cartelization, special interest legislation, and favoritism in policy administration. Swedish agricultural policy exhibits these tendencies to a degree, but for a convincing case to be made that policy has been dominated by an iron triangle, there would have to be evidence of at least one of the following conditions:

- Policy execution, where the triangle exerts the most influence, has greater impact on the farm economy than legislation, budget decisions, and price negotiations, where the triangle is weaker
- Farm organizations and the bureaucracy are able to translate their near monopoly in policy execution, via some kind of backward linkage, into leverage over farm legislation, budgets, and price setting as well
- The farm bureaucracy uses its relative autonomy to distort policy in favor of farmers when official objectives conflict with farmers' interests.

The farm organizations' collaboration with state agencies does confer political influence beyond farmers' numbers or their economic significance, yet the evidence does not persuade us that any of the three conditions for hegemony holds. Still, the Social Democratic government saw the agricultural bureaucracy as a conservative and even obstructive force in the 1980s, especially when administrators dragged their feet in responding to requests for reform proposals, including a plan to eliminate excess production and export subsidies permanently. This reinforced the leading politicians' sense that only a thorough policy overhaul would be effective and that the agricultural bureaucracy should be kept at arm's length from the process.

5.6 From Inertia to Reform

The preceding analysis supports the thesis that complex governance regimes are overdetermined by economic, political, and cultural-historical forces. Jointly, they create a bias toward the status quo, with policy failures and changing conditions prompting incremental reactions that tend to increase the policy regime's legal and administrative complexity without resolving its underlying contradictions. In Chapter 4 we described several such piecemeal modifications in the 1980s. With hindsight, we see that they were early signs of an emerging consensus among three key political parties—the Social Democrats, Liberals, and Moderates—that the entire interventionist superstructure needed overhaul. Three bellwether changes were:

- Rewriting the ground rules for price negotiations by eliminating the income parity formula
- Requiring producers to bear most of the cost of dumping surplus commodities on world markets

- Granting environmental rather than agricultural agencies authority to design and enforce environmental policies in the farm sector.

A central thesis of Petit's (1985) historical studies is that agricultural policy regimes do not collapse under their own weight. A decisive break requires convergence between agriculture's internal problems and critical conditions in the larger political economy. Our sense is that three broad socioeconomic conditions converged with agricultural policy failures to propel agriculture to the center of Sweden's political agenda in 1989–90. The first was structural and macroeconomic problems, the second was shifting issue coalitions among the political parties, and the third was a changing political culture.

The interventionist agricultural policy was inconsistent with the macroeconomic and structural reform measures enacted by the Social Democrats after their return to power in 1982. SAP pragmatists, led by Finance Minister Feldt, were able to impose a restrictive fiscal policy, deregulate the financial sector, negotiate a supply-side tax reform, and launch a range of experiments to combat the "cost disease" of public services such as child and health care. Indeed, under the influence of the Feldt faction, the Social Democrats in the 1980s embraced much of the neoliberal economic program of their bourgeois opponents. Proponents of this ideologically unpalatable medicine asserted that it was needed to reestablish the efficiency and dynamism on which Sweden's international economic competitiveness, private prosperity, and public revenues all depend (Jessop et al. 1991, Lindqvist 1991). Meanwhile, critics on the left interpreted the sweeping policy revisions, which coincided with a major corporate offensive against labor and the state, as a betrayal of the social democratic vision and as "the demise of the Swedish model" (Pestoff 1991).[14]

Since SAP leaders had long sought a pretext to reform agricultural policy, it was logical to extend their neoliberal reforms to agriculture when the chance arose. Export subsidies were a convenient target for budget cutters, and negotiated food prices became a prime focus when inflation reached double digits. With the national unemployment rate below 2 percent, the old postwar criticism that distortion of farm incentives stifled the exodus of workers from agriculture was once again mobilized. Apart from these domestic issues, the government

[14] Three key elements in Swedish corporations' strategic offensive were the Swedish Employers' Federation's rejection of centralized wage bargaining, a sevenfold increase in overseas investment in the 1980s, and a massive promarket, pro–individual choice, anti–state media campaign launched by business-financed organizations (Pestoff 1991).

sought to enhance Sweden's bargaining leverage in the GATT negotiations through a willingness to reduce agricultural trade distortions.

Several changes in the complexion of party politics toward the end of the 1980s facilitated the formation of a multiparty coalition for agricultural policy reform. Soon after its 1988 election success, the SAP found its support in the polls slipping badly. The attack it leveled at farm policy's inflationary and environmental defects can be interpreted as a tactic to win back voter support on a basic bread-and-butter issue and on the voters' top priority concern, environmental protection.[15] The Left party, a traditional Social Democratic ally, opposed most of the SAP's neoliberal reforms, so the minority government looked to the Liberal party for backing on economic issues. The Liberals had long shared the SAP's desire for a new approach to agriculture. Meanwhile, the Center party's voter support and parliamentary influence faded after its ineffective performance as a partner in the bourgeois government of 1976–82. With their rural constituency now a minuscule fraction of the electorate, Center leaders shaped a new program less firmly wedded to the old farm policy and more oriented to urban and suburban voters.

The parties and the mass media made sure that Swedish voters were informed of the old policy's poor goal attainment and cost-effectiveness. But at a deeper level, the political culture—collective consciousness and ethical commitments—that undergirded the Swedish model was also eroding. Stagnant incomes increased the preoccupation with personal material conditions. Global cultural homogenization ate away at national cultural values and moral traditions. And a generation of Swedes came of age who took affluence, material security, and public services for granted and had only a vague sense of the collective struggles and sacrifices that had been required to achieve them. This evolution in political culture lies behind shifting party allegiances and, specifically, waning public support for the old agricultural policy.

Objective conditions and subjective trends were mutually reinforcing. The legitimacy of price and income supports for a single economic sector is likely to be questioned by wage earners who have seen their incomes eroded by food price inflation and who no longer consider wage solidarity to be a bedrock principle. And people distressed by

[15] The Social Democrats received 43 percent of the vote in the September 1988 election. By June 1990, when the food policy reform legislation was enacted, their support was down to 33 percent in voter opinion surveys (Bergström 1991).

agriculture's detrimental effects on the environment are also likely to lose some sympathy for farmers operating industrialized production systems. In sum, by 1988 convergent forces inside and outside Swedish agriculture had transformed policy inertia into a momentum for change.

6

Greening: New Demands on Swedish Agricultural Policy

The environmental contradictions of Sweden's industrialized food system came into focus in the 1980s as a central aspect of the agricultural policy crisis and also as a major influence in the overall greening of Swedish politics. Greening opened the agricultural policy process to new actors and forced the traditional actors to modify their behavior. In policy decisions, greening meant a greater weight placed on values and ethical concerns regarding environmental protection, sustainable resource use, cultural-historic landscapes, wildlife habitat, human health, and farm animals' well-being. Environmentalism also became interwoven with the sociocultural values attached to family farming and rural community life. In sum, greening became a critical force in the broad movement for agricultural policy reform.

Agriculture provides a point of entry for indicating how greening reflects a new configuration of social forces taking shape in Sweden. This configuration is characterized by the uneasy coexistence of two powerful tendencies: ideological greening, with its emphasis on collective values, market failure, and political intervention; and neoliberalism, with its emphasis on individual choice, market orientation, and deregulation. In the following narrative we explore a pattern of tensions between self-interest and moral commitment, expediency and genuine conviction, materialist and ecological values, and rationalist and romantic attitudes toward nature and farming.

6.1 Mainstreaming Environmental Issues

The politicization of agroenvironmental problems is not a new phenomenon in Sweden. For a century or more, objections have been raised about the envionmental impact of certain farming practices and demands have been made for corrective state actions. Thus, in the late nineteenth century, conservationists decried the state-supported drainage of fragile and biologically rich wetlands which accompanied agricultural commercialization. Early in the present century, there was a movement to restrict windbreak removal, bare fallow, and other practices that worsened wind erosion on the southern plains. After World War II, conservationists opposed the conversion of meadows to row crops and marginal cropland back to spruce plantation when rationalization measures encouraged these practices. And in the 1960s and 1970s, an environmental coalition inspired by Rachel Carson's work forced a ban on mercury-based fungicides and then DDT. These examples are indicative of two strands in Swedish environmental policy in the years before a National Environmental Protection Board was created in 1967: land-use regulation and restriction of chemical pollutants.

Contemporary Swedish environmental activism, with its intense scrutiny of agriculture, is grounded in these precedents and traditions of media exposure and public awareness. According to Lars Lundgren (1989), it has even deeper roots in a Protestant-romantic construction of nature.

Notwithstanding the precedents, past agricultural policy had no systematic environmental component. In fact, the 1960s agricultural commission forged a consensus between agroindustrial interests and wage earner–consumer interests to insulate farmers from excessive environmental restrictions. The 1967 agricultural legislation stated, "Environmental conservation questions will not be handled in agricultural policy, but rather in environmental protection policy" (quoted in Lundgren 1989:68). Under this division of labor, agricultural statutes such as the Land Care Law (1979) were designed not to protect landscape values or biological diversity but rather to maintain Sweden's 2.9 million hectares of farmland for national food security and for regional economic development purposes (Drake and Petrini 1985).[1]

In the 1970s and early 1980s, a few environmental organizations kept up a persistent critique of chemical-intensive food production, and an alternative agricultural movement was taking shape, but mainstream envionmental organizations and the mass media did not empha-

[1] To agroindustrial interests, the Land Care Law is one means of propping up the agricultural economy in disadvantaged farming regions.

Figure 6.1. Forces and actors in the greening of agricultural policy

a. *Impacts and perceptions of negative environmental symptoms*

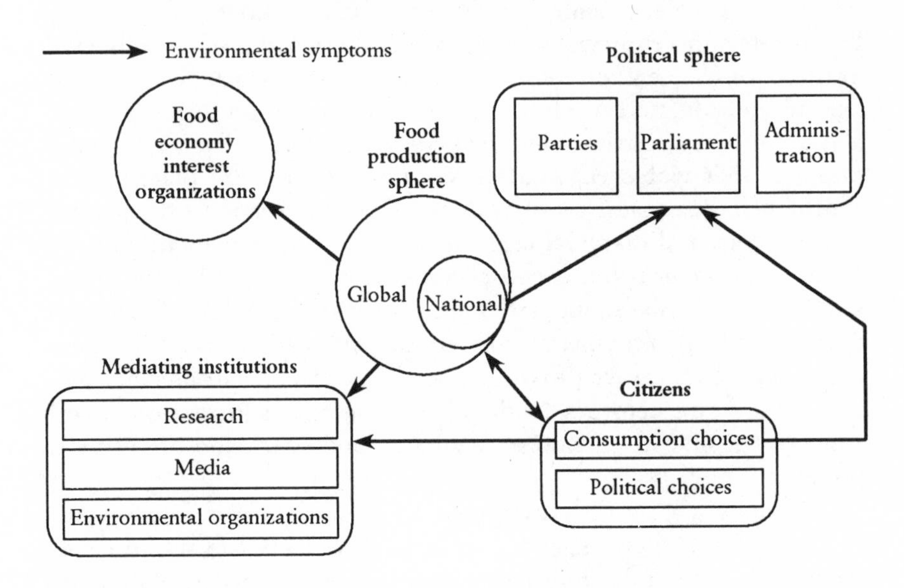

b. *Actors and interactions*

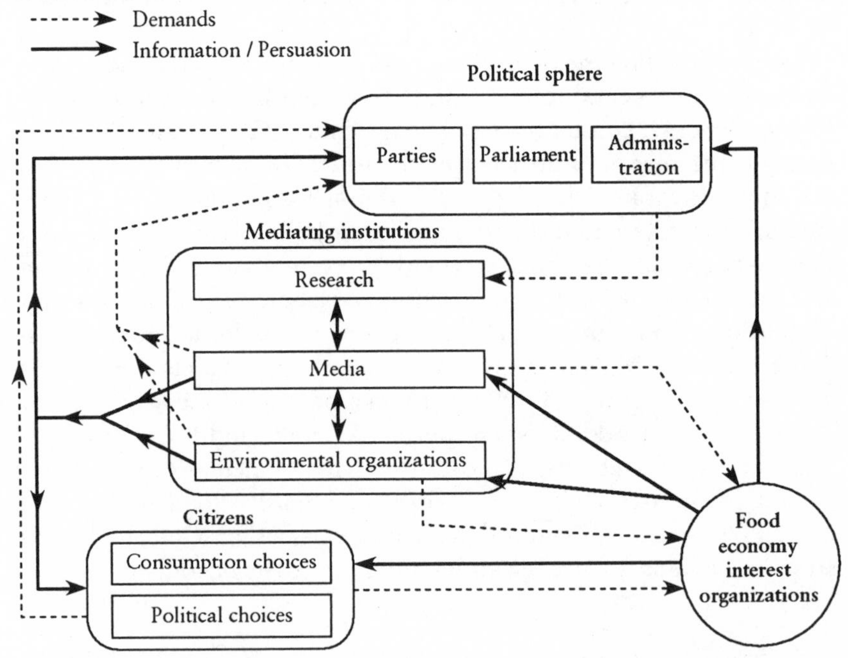

size agriculture's environmental effects (Gillberg 1970).[2] These were years of anxiety about food security, and the policy debate focused on ensuring food self-sufficiency by stimulating production. (Recall that the degree of national self-sufficiency had declined under the low-price policy of the 1960s and early 1970s.) In the mid-1970s, the decision was made to increase farm prices sharply and the County Agricultural Boards intervened to prevent farmland abandonment or conversion. These moves paralleled the EC's stimulation of agricultural investment and the United States's "hedgerow to hedgerow" planting policy.

Early in the 1980s, a long list of agroenvironmental problems surfaced in the media and the policy arena, just as the food supply-demand balance was swinging unexpectedly from scarcity to glut. The greening of the policy debate reflects a spreading awareness that agricultural policy itself was causing or at least intensifying adverse side effects (enumerated in section 6.2). This recognition strengthened demands that environmental protection be added to the goal triad of food security, distributional equity, and reasonable consumer prices. The 1985 food policy legislation reversed the 1967 exclusion of environmental problems from the agricultural policy domain by incorporating an environmental action plan. A cluster of three goals—environmental protection, nature conservation, and sustainable agriculture—became Sweden's fourth broad agricultural policy objective.

Figure 6.1 depicts the major actors and central forces in the greening of Swedish politics. The arrows indicate the pathways by which adverse environmental symptoms in the food production sphere have been interpreted and acted on by mediating institutions. These institutions are directly involved in politics, but they also transmit information and analysis to citizens, shaping their expectations and demands. The greening of Swedish political life has taken place through many channels that link state institutions (parliament, cabinet, government agencies, political parties) with the mass media, public interest environmental organizations, and scientific-technical institutions. Economic interest organizations have made reactive efforts to clean up

[2] A qualification to this generalization was the recurrence of proposals for development of agricultural bioenergy after the 1973 oil shock. Advocates saw this as a way to reduce Swedish dependence on nuclear power, whose elimination was their top priority. Bioenergy supplies might also limit dependence on fossil fuel imports and reduce pressure to dam Sweden's four remaining wild rivers for hydropower. State-supported bioenergy development has been a cornerstone of LRF proposals in recent years (see Chapters 8 and 9).

their public image, and all types of political actors have attempted to shape public opinion.[3]

A simple diagram cannot display all the instruments employed by the actors: news reportage, advertising, opinion polls, public demonstrations, scientific reports, legislative lobbying, and election campaigns, to mention a few. Neither Figure 6.1 nor the discussion that follows does justice to this complexity: it is only a crude approximation of Swedish political reality.

Andrew Jamison (1987) traces the present phase of Swedish environmental politics back to 1980, when voters in a national referendum supported early decommissioning of Sweden's nuclear power plants. By this interpretation, a key feature of the current phase is the environmental organizations' changing status, priorities, and methods. The largest membership organization, the Swedish Society for Nature Conservation (SNF), was fairly apolitical in the past. But it has taken an irreversible political step beyond its main historic function of promoting nature appreciation. SNF's leaders have been put on the spot to adopt a more decisive political stance by an increasingly politicized membership and especially by an idealistic youth organization, the Field Biologists. Indeed, as a new generation of former Field Biologists has become SNF's leaders, political activism has come to seem natural.

Coming from the opposite historical direction, several grass-roots environmental groups that were linked to the radical social movements of the 1960s and 1970s have muted their radicalism and revised their strategy of working outside conventional political channels. Most have evolved into more highly structured, professionally run organizations with agendas for reform, rather than transformation. Growing popular support for the Swedish Green party (Miljöpartiet de Gröna) and for Greenpeace was a prominent sign that environmentalism was becoming a mainstream cause. The Green party emphasizes participatory grass-roots organization, but its local and later national electoral successes in the late 1980s required scores of party members to participate in corporatist governance processes in which pragmatic coalition building and compromise were the norms. Meanwhile, Greenpeace

[3] Individuals have multiple and often contradictory roles that cannot be shown in Figure 6.1. Beyond their private choices as consumers, workers, and voters, they also participate in the institutions shown in the figure. Business lobbyists, journalists, environmental activists, scientists, bureaucrats, politicians: all are socialized to particular values and behavioral norms by their institutional roles and experiences. At the same time, they bring to these roles their own material interests, values, and convictions. To take a single example, the work of a typical Swedish journalist is affected by the fact that she also belongs to both a political party and at least one environmental organization.

Table 6.1. Voter priorities in election years, 1976–1988

Year	Ranking of environmental concern (among 12 issues)	Percentage of voters ranking environment in top 3 concerns
1976	3	33
1979	4	32
1982	5	32
1985	2	41
1988	1	56

Source: SIFO 1989:27.

gained widespread public respect and financial support in Sweden (as in much of Europe) for its tenacious media-oriented campaigns. Greenpeace Sweden is essentially a professional rather than mass-based or democratic organization (Jamison 1987).

As environmental concerns and issues became part of mainstream politics, there was an inevitable shift of emphasis from the philosophical toward the practical, from idealistic purity toward workable compromises, and from holistic visions toward discrete measures.

A second commentator, Lars Lundgren (1989), dates the current phase of environmental politics from 1985 and views Swedish politics in the preceding decade as dominated by the macroeconomic crisis described in Chapter 2. He sees nuclear power as the only galvanizing environmental issue in the early 1980s. The year 1985 may indeed have been a turning point. Before 1985, environmental protection ranked fairly low among voters' concerns in election years (Table 6.1). In 1986, the year of Chernobyl and North Sea fish kills, environmental protection surpassed full employment for the first time as the top-priority policy concern in citizens' minds. By 1987, it ranked first among supporters of every political party across the left-right spectrum. In the 1988 election, environmental protection was rated the most important issue by over half the voters, and every political party made a special effort to appear environmentally friendly. The Greens took 5.5 percent of the vote to become the first new party to gain Riksdag representation in more than seventy years.[4] An indication that environmentalism is not merely a political fad in times of full employment and prosperity is that a plurality of voters listed environmental protection as their top priority during the 1991 election cam-

[4] In Sweden's system of proportional representation, a party must receive 4 percent of the vote in a national election or 12 percent of the vote in a municipal election to be seated in the Riksdag.

paign, despite Sweden's rampant inflation, rising unemployment, and deep uncertainty about political stability (Sjöström 1989, Taylor 1991).

As in other industrial nations, the greening of public opinion in Sweden reflects a convergence and interweaving of several forces. Five main influences can be identified:

- Long-term increases in economic productivity and material affluence
- Dramatic environmental disasters and ominous predictions about the future of the ecosphere
- An explosion of knowledge, scientific and popular, about the environment
- Instant global communication of news about environmental events
- The growing political effectiveness of environmental advocacy organizations.

Increased economic productivity and material affluence affect both the demand for and potential supply of environmental services. Swedish material living standards have been high and secure, and paid labor has fallen to fifteen hundred hours per year. There is a positive income elasticity of demand for environmental amenities. On the supply side, the same production system that underlies mass consumption and the attendant environmental degradation is creating the technological capacity to mitigate environmental problems and supply environmental amenities.

Green merchandising in Sweden is a recent twist on the inherent conflict between mass consumption and environmental protection. It is problematic whether producers' efforts to boost sales and profits by promoting environmentally friendly commodities should be thought of as responses to an autonomous greening of consumers' preferences or as a manipulation of those preferences. Still, the fact that most Swedish consumers are willing to pay a premium for low-phosphate detergent, unbleached and recycled paper products, and meat from animals raised in humane conditions reveals a moral dimension in their behavior.[5]

The upsurge in Swedish environmental concern in the latter 1980s took place in a setting of mixed economic phenomena. Full employment and a declining share of workers in highly polluting industries

[5] If consumers were pure utility maximizers, they would pay more for environmentally friendly products only if they were "buying" a good conscience. Since their individual material benefit from paying more would be negligible, their incentive would be to free ride on environmental benefits provided by others.

probably weakened opposition to tougher environmental standards. But average labor productivity and real household income were nearly stagnant. Even more perplexing, the greening of public opinion coincided with a surge of consumer spending financed by unprecedented borrowing: that is, a choice of the present over the future. The apparent inconsistency between conservation ideology and materialistic behavior casts doubt on any claim that Swedish society had entered a postmaterialist age (Olingsberg 1990).

Dramatic environmental disasters, signs of cumulative environmental deterioration, and predictions of future disruptions on a massive scale were a second set of influences. They operate at every scale from the local to the global. A powerful national example is *bilism* (car culture), with its adverse effects on ecology (congestion, land appropriation, local air pollution, greenhouse gas emissions, and so on) and economy (energy waste and import dependence). Citizens are coming to realize that the pervasive dependence of Swedish production and consumption patterns on fossil fuels is a central cause of environmental degradation and a primary reason why present environmental problems will be far more difficult and costly to solve than the discrete pollutants dealt with in earlier phases.

Awareness of Sweden's vulnerability to external forces has been reinforced by the lingering radiation effects of Chernobyl, the flow of toxic effluents into the Baltic Sea from former Soviet republics, and the acidification of eighteen thousand Swedish lakes, mostly by airborne pollutants from abroad. All the global environmental phenomena that dominated news in the latter 1980s—ozone depletion, climate change, and rising sea levels—could have enormous impact on Sweden, and not least on Swedish agriculture.

Most Swedes remain fairly optimistic about solving discrete domestic environmental problems, a tribute to the nation's past record of technological ingenuity and fairly effective policies. But when Svensson ponders the global scene and the longer run, he is increasingly skeptical about the sustainability of the present industrial mode of production and the consumption standards it has supported (SIFO 1989b).

Instant transmission of environmental news is a third and closely related politicizing force. As citizens of the "global village," Swedes are as aware of devastating floods in Bangladesh as of threats to aquatic life in nearby lakes. Although few grasp the complexities of a phenomenon such as the degradation of aquatic ecosystems, Swedes are comparatively well informed. Nearly two-thirds of adults, for example, read daily newspapers, versus about one-fourth in the United

States. Newspapers are state-subsidized and continue to rival television as the prime source of popular knowledge about the environment (SIFO 1989b).

It has been argued that the mass media were slow to develop environmental reporting in the 1960s and that their emphasis on socioeconomic problems in the 1970s deflected public attention from mounting environmental problems (Thelander and Lundgren 1989). But the print and electronic media jumped onto the environmental bandwagon in the 1980s, and some would claim that they drive it. More or less objective analysis is intertwined with editorializing and debate articles, and news coverage blends horror stories about catastrophes with testimonials about solutions.

The mass media's role in political discourse is complex. Media institutions and prominent personalities within them are relatively autonomous actors in their own right. Indeed, within limits, editorial decisions make the news. Yet the media are also an arena in which other actors compete for favorable coverage of their positions. Figure 6.1b sketches the "input-output" relationship involved in the production of news: the media selectively filter, focus, and amplify input from the "outside" before transmitting it back again. (Petersson and Carlberg 1990).[6] We note in passing that 25 percent of Swedish journalists belong to at least one environmental organization, compared to 8 percent of citizens at large (Thelander and Lundgren 1989).

A fourth force, linked to the previous two, is the explosion of scientific and popular knowledge about the environment. Sweden's Environmental Protection Board, for example, has presided over an exponentially expanding research budget; in its quarter century of existence, the annual production of technical and policy studies has increased nearly tenfold. Findings from publicly funded research are publicized by the media and converted to new demands by environmental activists. These in turn shape public opinion and political decisions, leading to a still greater resource commitment to scientific investigations (Figure 6.1b). As a result, both policymakers and ordinary citizens have more systematic knowledge and sheer empirical information about the ecosphere than ever before.

The worldwide knowledge explosion is exemplified by the growing capacity to monitor, model, and predict global climatic conditions. More than a century ago, the Swedish scientist Svante Arrhenius

[6] We are not qualified to analyze the internal workings of media institutions, competition in the media industry, or state-media relations, all of which affect editorial slant and the priority accorded potential news items.

speculated about a greenhouse effect caused by atmospheric concentration of carbon dioxide from fossil fuel combustion. But the capacity to measure changes in atmospheric conditions and simulate enormously complex climatic interactions (however imperfectly) requires technical capabilities in satellite monitoring, computation, and global circulation modeling that did not exist even a decade ago. Human intuition and wisdom may not have progressed since Arrhenius's time, but knowledge has expanded phenomenally.

A fifth force, discussed previously, is the environmental movement's increased political sophistication and influence, based on growing professionalism, expanded membership, and successful fundraising. (Use of the term "movement" does not imply political unity.) At one level, environmental organizations have become adept at using news of catastrophes to mobilize new members, draw media attention, and create election issues. Thus in 1987–88, a joint public relations campaign by SNF, Greenpeace, and the Green party elevated public anxiety about seal deaths and toxic algal blooms on the Swedish West Coast to such a level that the 1988 election campaign was driven in a green direction and the Green party won seats in the Riksdag (Micheletti 1989).

At another level, members of environmental organizations have been legitimated as participants in corporatist processes, ranging from legislative commissions to technical advisory panels. SNF is now routinely asked to second experts to government investigations, including those dealing with aricultural pollutants, farmland use, and animal husbandry. In a green version of the "revolving door" between private and public employment, former environmental activists (most conspicuously former Field Biologists) now staff key positions in several state agencies. In fact, when the Ministry of Environment and Energy was established in 1987, it would have been impossible to fill its technical and managerial ranks competently without tapping former activists. Now that many top-level political leaders publicize their SNF membership, environmentalists have ceased to be "them" and have become "us" (Jamison 1987).

SNF, a broad-based membership organization with strong establishment credentials, and Greenpeace, a cadre-based organization with a history of militant action, both espouse visions of a postindustrial society. Yet both capitalize politically on their reputations as reasonable, uncorruptible spokespeople for the public interest, especially when it stands in opposition to the short-term interests of capital, labor, political parties, and the bureaucracy. The media generally have a high regard for the rigor and honesty of environmental organiza-

Table 6.2. Credibility of Swedish sources of environmental information, 1988

Percentage of survey respondents who usually believe statements by:	
Researchers	89
Environmental groups	73
Television reporters	58
Economic interest groups	25
Politicians	20

Source: SIFO 1989b:10.

tions' analysis, and a recent opinion survey suggests that this trust is shared by the public (Table 6.2).

Responding to the five forces described in the preceding pages, political parties and economic interest groups of every stripe profess concern for the environment. Until a few years ago, labor unions and industry organizations, with allies in the Social Democratic and Moderate parties, respectively, tended to put a one-sided emphasis on the costs of environmental protection: stagnant productivity, erosion of international competitiveness, job loss, deterioration of industrial regions, and declining living standards. Such arguments have not disappeared; indeed, they have been reasserted in the current recessionary period. But, characteristic of the negotiated economy, organized labor and capital have joined the search for environmentally friendly solutions to such dilemmas as dependence on automobile transportation and fossil fuels. Clearly, this is partly political opportunism: environmental insensitivity would provoke public ill-will. Often it also has direct economic benefits: strong environmental policies create new economic opportunities and new forms of public subsidy.[7] But we submit that the greening of labor's and capital's behavior also reflects the development of a genuine ethical commitment on their part.

When the voters' green priorities and the Green party's growing support became clear in 1988, Prime Minister Ingvar Carlsson reoriented the Social Democratic election campaign to shed the SAP's image as the "concrete party." Claiming credit for a wide range of new green laws and programs, most of which were in fact multiparty efforts, he

[7] Michael Porter observes in *The Competitive Advantage of Nations* (1990:141) that Swedish corporations have effectively turned the country's tough environmental standards into export opportunities. There are growing international market niches for environmentally friendly Swedish products, such as chlorine-free white paper, and for cutting-edge pollution-abatement technologies, such as electrostatic precipitators and nuclear waste storage systems.

sought to persuade voters that the SAP was Sweden's "true environmental party." (American readers will recognize a parallel with the 1988 U.S. election campaign, in which candidate George Bush launched a green offensive to maintain the political status quo by proclaiming himself the "environmental president" in spite of his party's abysmal record on environmental protection.)

Notwithstanding the conservative shift in Swedish economic policy in the 1980s and the present preoccupation with severe macroeconomic problems, environmental friendliness has become a litmus test for all political actors. The same appears to be true in every advanced industrial nation. An analysis of the interests and past behavior of political parties and economic interest organizations justifies suspicion that they are playing an opportunistic game with the environment (see section 6.3). Nonetheless, the momentum behind environmentalism in the latter 1980s revealed the "civilizing power of hypocrisy" mentioned in Chapter 5: political players were forced to back their words with actions or risk a loss of legitimacy. At the same time, green political pressure often induced hasty or largely symbolic political reactions with little tangible environmental benefit or with environmental improvement bought at an exorbitant cost.

6.2 The Agricultural Connections Come into Focus

Shortly after environmental protection was made an official agricultural policy goal in 1985, two statements that were developed outside the agricultural establishment signaled that green issues would be high on the policy agenda. The National Environmental Protection Board's 1986 report *Agriculture and the Environment* identified three main conflicts between modern farming methods and the goals of environmental protection and nature conservation: nutrient leaching into ground and surface waters, contamination by chemicals and heavy metals, and destruction of habitat for endangered flora and fauna. Nitrogen leaching (over 48,000 metric tons per year) was considered the most urgent problem; cadmium deposition (6.3 mt/year), the cumulative loss of one-third of Sweden's biologically rich meadows and pastures since 1945, and degradation of arable landscapes were also sources of concern (NEPB 1986, SS 1991c).

The following year, the newly created Ministry of Environment and Energy submitted the government's broad environmental policy proposition, *Environmental Policy on the Eve of the 1990s*. Agriculture's environmental deficiencies featured prominently alongside energy and transportation problems, and another green issue was added to the debate: farm animals' well-being (MOE 1988).

These documents challenged two deep-seated traditions: the presumption that farming is synonymous with resource stewardship and nature conservation, and the autonomy of corporatist agricultural decisionmaking. In effect, the greening of agricultural policy precipitated turf battles within the state apparatus, one between the Ministry of Agriculture and the Ministry of Environment and Energy and another between the National Board of Agriculture and the Environmental Protection Board.

As previously indicated, agriculture's negative environmental effects had several sources, including market forces, lack of knowledge, technological developments, tax laws, and farm policies. Economic pressures and incentives increasingly pitted the farmers' individual rationality against collective rationality. The problems that most clearly derived from policy failures were:

- Tax distortions—heavy on labor and light on capital—strengthened the bias toward mechanization. This in turn contributed to larger field scale, soil compaction, and removal of such field impediments as stone walls and tree islands.
- Commodity price supports stepped up the farmer's optimum fertilizer and agricultural chemical applications. They were increasingly used as substitutes for crop rotation and cultural practices to maintain soil fertility and control pests and weeds.
- Artificially high crop prices strengthened economic incentives to convert species-rich meadows and pastures into cropped fields. Selective price supports favoring grain and oilseeds reinforced the trend toward specialization and simple rotations, with the effects noted above.
- The evolution toward monoculture encouraged greater use of herbicides and fungicides.
- Rationalization measures deliberately encouraged larger farm scale, specialization, mechanization, wetland drainage, and elimination of field impediments.
- Rationalization measures, coupled with high labor costs, encouraged factory-like livestock confinement systems that were inhumane to animals and worsened manure management problems.
- Farmer training and extension programs promoted economistic attitudes about agriculture as a cost-minimizing, profit-maximizing business. They directly encouraged chemical-intensive soil and pest management.

Although the media continued to cast farmers in a generally sympathetic light, the amount of reporting about modern farming's negative effects grew sharply. Newspaper articles whose central topic was agriculture's economic difficulties, such as overproduction or food price inflation, increasingly touched on related environmental concerns, such as excessive fertilization and chemical residues in food. Reporting on alternative agriculture, in contrast, was typically cast in a testimonial mold: positive, unproblematic, and sometimes romanticized (Pettersson 1989, Westberg 1988). The media—like policymakers—tended to highlight environmental impacts in piecemeal fashion, although more comprehensive critiques of the agroindustrial complex occasionally appeared, along with speculations about what an environmentally sustainable alternative might look like (cf. Michélsen 1987a, b). The environmental emphasis in the media's treatment of agriculture undoubtedly helped to make farm policy a "front burner" political issue; it also blunted the iron triangle's capacity to shape the policy agenda.

Rising Green Expectations

Green demands on farmers are not limited to reducing negative externalities but also include supplying public goods. Agriculture in the past was expected to strengthen national security and regional development; the affirmative emphasis now is shifting to maintenance of valued landscapes and biotopes, safe recycling of municipal wastes, and production of renewable biofuels.

In the 1980s, up to 60 percent of municipal sewage sludge was spread on agricultural land.[8] The figure plummeted to less than 10 percent when the LRF imposed a boycott on sludge applications, motivated by fear of adverse consumer reaction to the risk of heavy metal contamination and by a desire to reaffirm farmers' reputation as environmental stewards. LRF has since ended the boycott in response to widespread public support for waste recycling and the development of methods of reducing cadmium and other contaminants (Hedvåg 1990, NEPB 1989).

The age-old Swedish agricultural tradition of producing heat from farming residues and farm woodlots has gained renewed interest for

[8] Thousands of tons of sewage sludge were applied to arable land each year before the environmental and economic effects had been thoroughly investigated. Indeed, environmental risk assessment and economic cost-benefit analysis related to sludge application only began to reach the level of sophistication needed for well-informed decisions in the early 1990s.

several reasons. By the mid-1980s, it had become obvious that any policy to eliminate the grain surplus would render a large area of arable land redundant; much of that land would be agronomically well suited to energy crops. Second, according to some analysts, technical research and development had brought bioenergy to the threshold of commercial viability. Third, Sweden's existing energy strategy was not tenable. The major policy objectives—decommissioning nuclear power plants, protecting the remaining wild rivers from hydroelectric development, reducing dependence on fossil fuel imports, and cutting air pollution—were mutually inconsistent. A 1988 Riksdag resolution to freeze carbon dioxide emissions, followed by the imposition of CO_2 and energy taxes in 1990, compounded the policy dilemma. Many environmentalists, cheered on by the LRF, began to look to agricultural biofuels as one way out of the bind. (Bolin et al. 1989, Kumm 1989, NEPB 1986). This prospect is assessed in Chapters 7, 8, and 9.

6.3 Playing Green Politics with Agriculture

The multiparty coalition that negotiated an agricultural policy reform at the end of the 1980s was principally concerned about the old policy's economic costs and irrationalities. But the agricultural situation also presented opportunities for coalition members to demonstrate their greenness to a demanding public. The parties and the agroindustrial interest groups acted as if approval from environmental organizations and the public really mattered.

Sweden's major environmental challenges, such as reducing automobile dependency and phasing out nuclear power, have no quick solutions; furthermore, any long-term resolution is bound to entail heavy socioeconomic costs and dislocations. These difficulties make it tempting for agents to demonstrate their environmental friendliness by playing green politics with agriculture. It offers several tractable challenges, such as cleaning up eutrophied fishing waters, reducing pesticide residues, and protecting highly valued landscapes. Each of these problems is highly symbolic, limited in scope, and remediable at a comparatively low cost, whether the cost is measured in state expenditure, consumer impact, or interest group antagonism.

Game theoretic notions help to clarify why all the agricultural policy players adopted a green stance and why numerous piecemeal green initiatives were taken in the last half of the 1980s. In the preceding chapter we construed political behavior in the negotiated economy as a blend of tactical opportunism and genuine moral conviction. Here we stress interests and tactics.

A few generalizations about green politics are in order. Some players color themselves green in an effort to stand out from their rivals, while others do it to blend in with them. This blending effect helps explain the multiparty consensus on numerous green initiatives preceding the 1988 election. Actors attempt to influence and also win support from the public—or key subgroups. For the political parties, this is essentially a zero-sum game, with some attempting to increase their voter share and others trying to minimize losses. Despite these efforts (or perhaps because the efforts are so transparent), politicians are not very credible to voters on environmental issues: their 20 percent credibility rating is below even that of economic interest groups (see Table 6.2).

The various parties obviously have unequal influence in actual policy decisions. The Green party's surging popularity in 1987–88 forced established parties to intensify their proenvironmental rhetoric and actions, and the Greens' electoral success led to predictions that they would be in a decisive position between the bourgeois and socialist blocks. Once in the Riksdag, however, the Greens had little influence in agricultural policy. Their uncompromising stance on contentious issues, such as advocating a total ban on biocides, was too extreme for the more pragmatic parties. But they also failed to establish themselves as an attractive ally on nonenvironmental issues, where collaboration might have meant greater leverage over environmental measures.

The Liberal party, although a strong proponent of nature conservation for decades, had not been prominent in agricultural policy. In the 1980s, however, it was the Liberals who raised the idea of taxing agricultural pollutants and paying farmers directly for supplying such public goods as biologically rich landscapes. These recommendations reflected the Liberals' preference for economic levers rather than "command and control" measures. In contrast to the Greens, the Liberals made progress with their green agenda by being flexible on the details. Despite their small share of Riksdag votes (12%), Liberals could influence environmental policy because the Social Democratic government needed their backing on its key economic proposals, such as the 1989 tax reform. (This alliance echoes the SAP-Agrarian party cow trade of an earlier epoch.)

Until the Green party's surge, the Center party had a reputation as Sweden's most staunchly proenvironment party. Although the Center supported most green agricultural initiatives, it generally opposed any measures that imposed costs on its farmer constituents. The Moderates were similarly constrained by the fact that large-scale farmers were members and also influential party leaders. Recall that the Center party's Torbjörn Fälldin, a commercial sheep farmer, was prime minister in the coalition governments of the late 1970s and early 1980s (Micheletti 1990).

Based on their voting records, the Social Democrats and Moderates are Sweden's least environmentally committed parties, yet both backed green agricultural measures. The SAP championed green initiatives at least in part to secure support for its economic objectives: lower farm export subsidies, deregulated domestic markets, and authorization to negotiate lower import duties in the GATT round. But the Social Democrats' environmental policy was described by environmental leaders as "much talk, little action." Indeed, when a food sector working group was formed in 1988 to design a new farm policy, the government deliberately excluded environmental interest groups (as well as consumer and producer representatives) (Ihse et al. 1990).

The National Farmers' Federation faced a complicated strategic dilemma. Its overarching challenge was to protect members' economic interests, which depended heavily on the existing policy. In advocating continued import protection and price supports, the LRF now had to counter charges that this policy undermined environmental protection, nature conservation, and humane livestock husbandry. These charges were in addition to the (by now familiar) economic arguments against the policy's excessive cost to consumers and taxpayers and its poor attainment of socioeconomic objectives. LRF leaders also faced internal divisions, notably a split on green issues such as pesticide restrictions and subsidized afforestation of arable land. Some ecologically minded growers and small part-time farmers seceded from the LRF, and such alternative agriculture organizations as the National Organic Farmers' Federation and the Small Farmers from the West began to challenge LRF's hegemony as the farmers' political voice (AERI 1988, Källander 1991, Micheletti 1990).

The LRF's reaction was a public relations offensive highlighted by green policy proposals. Among other things, the LRF:

- Categorically refused to allow further applications of municipal sewage sludge on farmland, so long as there was any risk of unsafe materials entering the food chain
- Proposed a cross-compliance scheme so that participants in the acreage

set-aside program would have to plant fall cover crops on leaching-prone land
- Espoused more humane livestock housing conditions, with the proviso that farmers be given ample lead time for conversion and be protected from cheap imports
- Encouraged a joint state-LRF effort to develop export markets for organic and reduced chemical foods
- Became the most ardent proponent of biofuels production on redundant arable land.

The key issue for farmers was who would pay for these measures. Reichelderfer points out the obvious: such measures "require capital investment, high adjustment costs, increased operating costs, or some forfeiture of yield, any of which can reduce a farmer's income, at least in the short run" (1990:202). LRF's strategy predictably stressed solutions that would operate through existing corporatist administrative arrangements, shore up farm revenues, and minimize the farmers' share of costs (Dockered 1989, LRF 1990a).

In the contest for favorable public opinion, the LRF intensified its media campaigns. Between 1985 and 1990, LRF's director, Bo Dockered, became a high-profile media personality, with a stream of newspaper debate articles and radio and television appearances designed to create a new image of family farmers as resource-conserving, environmentally friendly, and quality conscious, to put alongside their traditional image as stewards of the living countryside (Hellström 1991, Micheletti 1990).

As mentioned, the entry of new players into the agricultural policy game changed the process and made the outcome less deterministic than in the heyday of the iron triangle. Micheletti summarizes:

> The treatment of the environmental issue . . . contrasts greatly with the problem of surplus production. The basic difference is the absence of a well-established network of corporatist exchange for the involved actors. . . . The issue of surplus production became an agenda issue because of its cost to the state, farmers and consumers. The environmental issue grew in political importance due to public concern, which forced the state into action. Whereas the first issue was generated through the corporatist exchange network, the second one was initiated by forces outside the classic agricultural subgovernment. (1990:172)

Organizations espousing environmental protection, nature conservation, animal rights, and alternative agriculture have made new demands to which the mass media have given extensive and generally favorable coverage. As mentioned, environmental organizations now have representatives inside the policymaking process: they are formally invited to submit commentaries on proposed legislation and they second experts to investigative commissions. Their members also staff much of the growing environmental bureaucracy.

Facing both political risks and opportunities, the Social Democrats and the opposition parties agreed on a series of new measures with a high symbolic payoff and low or, better still, deferred costs. Thus, for example, for a mere US$2 million per year in "rental" payments, twenty thousand hectares of treasured *hagmark* were protected. And livestock-housing requirements, to ensure animals a more "natural" way of life, were written with a ten-year lead-in period.

Our accent on political expediency is not meant to negate the good economic sense of building lead time into the implementation of green measures. Clearly defined long-term standards create incentives to develop appropriate technology and cost-effective management systems and also give state agencies time to fine-tune regulatory details and put support systems in place. In the reality of the latter 1980s, however, green demands provoked policy responses that were not always very coherent or farsighted. In the rush toward greenness, complex interrelated problems were reduced to digestible "policy bites," and haste led to shoddy policy design. Lars Lundgren summarizes: "We often prioritize the short-sighted and urgent. Running errands rules out investigations, experimentation rules out supervision, and so on. One must have a strategy, more farsighted all-inclusive thinking, as a corrective to the obtrusive reality of the urgent. Otherwise we will not get a good balance between the short term and the long term, the very important and the less important, what *must* be done and what can wait" (1989:84–85). We assess the consequences in Chapter 7.

6.4 Self-interest and Moral Responsibility

Swedish political behavior cannot be fully explained as a strategic game based on rational self-interested behavior. Rather, Swedish agricultural and environmental policies support the proposition that collective action is shaped by subjective forces, especially a historically evolved political culture. This culture is characterized by widely shared meanings, such as the historical resonance of particular agrarian landscapes; values, such as national food security; and ethical norms, such as protection of biological diversity. Environmental ethics, always in the background of Swedish agricultural policy, shifted toward the foreground in the 1980s.

Participation in the processes of the negotiated economy inclines agents toward what in less cynical times was called the common good. In the following discussion, we presume that virtually all bureaucrats, politicians, and interest group agents share, in varying degree, the meanings, values, and ethical standards described above. Although our causal analysis emphasizes the interaction of broad structural forces and contingent events in explaining major policy changes, morally committed individuals sometimes play significant roles in mobilizing citizens and shaping the agenda. The beloved storyteller Astrid Lindgren has played such a role in the movement to improve farm animals' well-being and protect the open farm landscape.

The popular demand for environmentally friendly agriculture certainly reflects citizens' perceived self-interest. They recognize that their present and future well-being are enhanced by safe and sustainable food supplies, clean water, an open and varied landscape, and vital rural communities in sparsely populated regions. These interests have been translated into demands for reduced agricultural pollution and for nature conservation, the two main traditions in Swedish environmental policy. But these interests are also dependent on sustainable resource use, the major new thrust of Swedish environmentalism.

Although green demands are grounded in self-interest, they cannot be reduced to self-interest. The mix of selfish and altruistic motives is well illustrated by Swedish attitudes toward protection of an open and varied farm landscape. General opinion surveys and contingent valuation studies reveal a nearly universal awareness of and concern about the loss of farm landscape qualities and a substantial willingness to pay to protect them. Today this concern seems to be Swedes' strongest motive for supporting the farm economy (AERI 1988, 1992; Hasund 1991). Contingent valuation studies, proceeding from neoclassical economic assumptions about optimizing behavior, indicate that most citizens discriminate among types of landscape and would pay significantly more to maintain farmland that possesses greater biological diversity, scenic beauty, and cultural-historic associations. Estimates of willingness to pay are subject to conceptual and empirical limitations, yet two findings from a 1986 survey are suggestive: Swedes expressed a willingness to pay about US$150 per hectare per year (current dollars) to maintain the 1986 farmland base, and the value placed on *hagmark* was more than twice that of a uniform grain landscape. The aggregate willingness to pay extrapolated from survey responses—$450 million per year, or $55 per capita—was nearly one hundred times the amount actually budgeted for landscape protection at the time of the study (Drake 1987).[9]

The noncommodity use values of open and varied landscape are fairly straightforward. Landscape amenities are "normal goods": as living standards rise, demand increases. In Sweden, where access to these amenities is institutionalized through *allemansrätt* (the right of common access), they are by and large public goods. Rising labor productivity has been used in part to buy increased leisure time, which in Sweden is institutionalized through legislative standards for the work week, vacations, and retirement age. Since leisure and landscape amenities are complementary goods, increased leisure has also shifted the demand for farm landscape outward. Finally, the cumulative loss of beautiful, biologically rich, or culturally significant land to such incompatible uses as grain monoculture, spruce forest, and housing estates raises the marginal social benefit of the remaining open land. Implicit in this utilitarian logic is the notion that citizens are aware that desirable landscape features such as stone walls and tree islands are joint products of particular types of farm activity.

Interpreting these use values is complicated by the fact that benefits do not depend entirely on in situ consumption of landscape services. Many Swedes derive vicarious satisfaction from the landscape through televised and literary depictions and even from the sheer knowledge that "it's still there."

Beyond its current noncommodity benefits, farmland has an option value, which economists typically describe in terms of the expected present value of possible but uncertain future uses (future benefits are discounted relative to current use values). Prospective benefits range broadly, from the land's contribution to food self-sufficiency in an international crisis to undiscovered medicinal properties of wild flora found in a particular ecological niche.[10] These hypothetical benefits are unevenly distributed across different types of farmland.

[9] We have conducted or supervised several contingent valuation studies on land and water quality protection. We do not lean heavily on aggregate measures of willingness to pay or point estimates of the nonproduction value of specific landscape types. The confidence intervals around such estimates are wide. Nonetheless, the aggregate willingness to pay is significantly greater than zero, and differences in valuation for various landscape qualities are statistically significant. Further, in repeated surveys, only about 10 percent of respondents express a zero willingness to pay to protect farmland (see AERI 1992, Drake 1987).

[10] Strictly speaking, possible future benefits that depend on knowledge that does not yet exist are termed "quasi-option values."

If people place a value on passing resource use options to future generations, then an element of altruism—a bequest value—enters the calculus. In various ways, these subtle and complex utilitarian notions have influenced Swedish land-use policy for many decades, but only recently have they been subjected to formal analysis and estimation.

The ethical commitment to farmland preservation rests on several overlapping motivations, which Allan Randall (1987) has termed benevolence toward present and future members of one's community or society, sympathy for people (e.g., farmers) and nonhuman life forms, and a more generalized sense of environmental responsibility. The first two distinctions are closely related to Etzioni's "I–We" paradigm, mentioned in Chapter 1. Some neoclassical economists translate such altruistic motives into a utilitarian category, the existence value of a resource stock.[11]

Recent evidence confirms that Swedes' moral commitment to environmental protection, including protection of agrarian landscapes, grew stronger in the 1980s (AERI 1992). Unfortunately, the contingent valuation exercises described above were not designed to capture the altruistic element in people's willingness to pay or to show how it has evolved over time.[12] A skeptic could claim that the rising valuation placed on farm landscape simply results from better information about the opportunity cost of lost farmland, but survey results strongly suggest that most citizens feel some moral responsibility as well. For example, biodiversity has only a tenuous connection to one's own utility, yet it ranks highest among motives for farmland protection; and nearly all respondents express a willingness to pay to maintain open land and landscape elements in places they can never expect to visit in person (Drake 1987, Hasund 1991).

It seems to us that purely utilitarian interpretations of greening obscure the real motivational importance of rights and entitlements. Regarding intergenerational obligations, Daniel Bromley summarizes a nonutilitarian logic: "The interests of the future are only protected by an entitlement structure that gives present generations a *duty* to consider the interests of the future. Future generations thus obtain a correlated right" (1989: 181; emphasis in original). Swedish environmental

[11] Economic discourse on the valuation of nonmarketed environmental benefits is summarized in Freeman 1993, Johansson 1987, Kriström 1990, and Randall 1987. Neoclassical economics has not yet converged to a uniform conceptual framework or set of categories to analyze environmental resources.

[12] Johansson (1987) shows that, even in the most highly controlled survey setting, the proportion of total willingness to pay attributable to specific motives cannot be accurately measured.

policy has evolved toward articulation of such entitlement rights, not only for present and future Swedish citizens, but also for noncitizens, via Sweden's unilateral taxation of carbon and sulfur emissions, and even for nonhuman species, via animal-rights legislation.

Both a moral commitment and a touch of romanticism were manifested in the popular demand for more "natural" treatment of farm livestock. This demand has been translated into new regulations regarding feed rations, housing conditions, freedom of movement, socializing opportunities, and slaughter procedures. The legislation followed an impassioned animal-rights campaign, inspired by Astrid Lindgren, and a wave of media exposés about animals' poor physical and psychological health and the brutal slaughter methods used in the industrialized livestock sector. It was widely understood when the legislation was passed that more humane husbandry would have a cost to consumers.[13]

Table 6.3 summarizes our interpretation of citizens' primary and secondary motives for five types of greening: pollution abatement, landscape and nature conservation, sustainable resource use, human health and safety, and humane treatment of farm animals. The five types are not, of course, as neatly separable as they appear in the table.

The table suggests the blend of self-interest and moral conviction that lies behind each type of initiative. There is no implication that every Swede is morally committed to all five objectives or that all citizens hold convictions with equal fervor. It should also be kept in mind that people make choices and express opinions in multiple contexts. They may practice different modes of rationality—and engage in contradictory behaviors—in their multiple roles as consumers, taxpayers, economic interest group members, community members, voters, and so on. As mentioned, Swedish consumers went on a spending spree for cars, new housing, and electronic gadgets in the 1980s, while giving top political priority to environmental protection. This could be interpreted in neoclassical terms as unproblematic "constrained utility maximizing behavior": a trade-off between scarce material and environmental goods. We believe, however, that it reflects an unre-

[13] Citizens' conflicting motives—and limited information—regarding livestock treatment are reflected in a 1988 survey of consumer attitudes. While the majority of respondents held that better living conditions for cattle, pigs, and chickens is a "very important" social goal, only 12 percent indicated that information about such conditions was important in their purchasing decisions (price and taste weighed far more heavily),and only one in six were willing to pay 20 percent more for meat from animals raised under high ethical standards (Holm and Drake 1989).

Table 6.3. Citizens' values and convictions in greening

Motivation for policy action	Types of green policy measure				
	Pollution abatement	Landscape and nature conservation	Sustainable agriculture	Protection of human health	Humane animal husbandry
Self-interest					
Current use values					
Clean air and water	P		s	P	
Food access and price			s	s	
Food safety/quality	P		s[a]	P[a]	s[a]
Recreational access	s[b]	s	s[c]		
Visual amenity	s[b]	P	s[c]		s[d]
Cultural-historic associations		P	s[c]		
Option values					
Food security		s	P	s	
Energy security		s			
Potential economic value of species[e]	s	s	s		
Future recreation		P			
Moral responsibilities (existence values)					
To future generations	s[f]	P	P	s	
To farm workers	P			s	
To rural society		s	s	s[g]	
For biodiversity	P	P	s[c]		
To farm animals			s[h]		

Note: P = primary motivation, s = secondary motivation.

[a]The perception of some consumers that food from organic farms and humanely treated livestock is more nutritious has not been scientifically verified.

[b]Reduced use of agricultural chemicals and fertilizer enhances these qualities via more complex rotations and more diverse biotopes.

[c]The presumption here is that conversion to chemical-free production reduces the homogenization of farm landscapes through simplified rotations and elimination of field impediments; less chemical use also increases biotope diversity.

[d]It is presumed that more "natural" husbandry would also protect or expand aesthetically pleasing grazing landscapes.

[e]As mentioned in the text, this is technically a quasi-option value.

[f]Fewer chemical residues in groundwater may reduce birth defects and the incidence of certain cancers.

[g]Psychological health.

[h]Chemical-free agriculture is likely to mean increased reliance on grazing and higher-fiber livestock diets, as well as fewer confinement operations.

solved, and for many Swedes a vaguely understood, moral tension: between selfishness and altruism, between short-term individual rationality and farsighted collective rationality. Materialist and ecological values coexist uneasily in the Swedish psyche—and in public policy.

6.5 From Green Demands to Piecemeal Policy Responses

The wave of green agricultural measures enacted in Sweden in the latter half of the 1980s was overdetermined by forces inside and outside agriculture and by an evolving mix of self-interested and morally committed political behavior. In a narrow framing, green policy revisions can be viewed as reactions to worsening environmental symptoms within agriculture, symptoms exacerbated by the existing policy. The pressure for change was intensified by the conjunction of environmental and economic symptoms. But in a broader framing, agricultural greening was one moment in a universal politicization of the environment.

A host of green policy measures was appended to the existing policy regime between 1985 and 1990. They were not inspired by a coherent, farsighted vision of sustainable agriculture, and indeed, the agricultural establishment viewed greening as an exercise in fine-tuning policy, not transforming it. We next turn to a critical appraisal of piecemeal greening.

7

Piecemeal Responses to Green Demands

The half-century-old Swedish agricultural policy remained in place during the second half of the 1980s, with green measures essentially tacked on. It is not surprising therefore that inconsistencies cropped up and that environmental contradictions remained. Since new policy instruments were often chosen with haste, it is also no surprise that economic cost-effectiveness has neither been assiduously pursued nor often attained. On the other hand, the tendency to set target dates for compliance well in the future has created space to fine-tune instruments, develop improved technologies, and make on-farm adaptations.[1]

Policy in practice is not the same as policy on paper. Although Swedish farmers, farm organizations, and agricultural agencies all profess their commitment to an environmentally friendly agriculture, bureaucrats may not enforce the letter of the law and producers may not comply with it fully when new policies threaten their interests.

As we have indicated, Swedish agricultural policy included some environmental provisions long before 1985. The most basic were farmland-preservation measures to prevent abandonment, afforestation, and nonagricultural development. The first two phenomena were special sources of concern in the 1950s and 1960s (and again today); the goal in preventing them was to ensure national food security. The loss of farmland to development was taken most seriously in the 1970s,

[1] Allowing substantial lead time for policy implementation is an interesting tactic from a public choice perspective. The government is able to adopt a posture that improves its public image, while blunting hostile reactions from adversely affected groups, depriving opposing parties of political campaign issues, and putting off budgetary costs to the future.

and the main concern was irreversible, long-term resource loss. *Allemansrätten,* the right of common access to all land apart from gardens and fields with growing crops, has customarily guaranteed recreational use of farmland. Food safety legislation also has a long history, though its central purpose has been to ensure antiseptic hygienic standards rather than to eliminate farm chemical residues. In the 1960s and 1970s mercuric fungicides, DDT, a few other pesticides, livestock growth regulators, and several veterinary medical preparations were prohibited, while registration standards were set for other agricultural chemicals. Finally, several general environmental regulations apply to agriculture, such as restrictions on chemical hazards in the workplace.

Related changes have occurred in agricultural research, education, and extension. Research into less input-intensive and less polluting production systems has increased sharply, using dedicated revenues from fertilizer and pesticide taxes. In 1987/88, 12 percent of state support for agricultural research was labeled "environmental."[2] Extension specialists are being trained or reschooled in such fields as integrated pest management, catch crops and green manures, humane animal husbandry, and organic farming. New courses on these topics have been introduced in the agricultural high schools and at the Swedish University of Agricultural Sciences. In sum, resource conservation and environmental protection are increasingly stressed throughout the agricultural training and extension systems; these programs complement the policy measures discussed here.

In keeping with the piecemeal character of green initiatives, this chapter considers a long list of measures, organized under four headings. Many instruments have multiple objectives and effects, however, preventing a perfectly consistent classification. The four categories overlap, and nearly all the measures contribute in some way to a more "sustainable" agriculture. Several measures were officially justified on environmental grounds but politically motivated by their contribution to reducing excess production, a reminder that the underlying policy regime was still defective. The four broad categories are:

- Reduction of agricultural pollutants to protect human health and ecosystems
- Preservation of farmland for its landscape amenities, cultural-historic values, and biological diversity
- Encouragement of more sustainable resource use

[2] Total revenue from fertilizer and pesticide taxes was roughly US$77 million in 1988 (NAMB 1988b, NBA 1989a).

- Articulation of farm animals' rights and improvement of their well-being.

A policy can be evaluated according to many criteria, such as its clarity of goal articulation, the internal consistency of instruments, enforceability, equity effects, goal attainment, and cost-effectiveness. In the case of environmental policies, consistency with the Polluter Pays Principle (PPP) for negative externalities and the Producer Compensation Principle (PCP) for provision of public goods is also relevant. We refer to each of these criteria below but primarily emphasize goal attainment and economic cost-effectiveness. Firm judgments cannot be made about many instruments, since relatively little analysis preceded their adoption and since their implementation did not include systematic performance evaluation. Of necessity, then, most judgments offered here are provisional, based on a mix of deductive reasoning and partial evidence.

7.1 Reducing Agricultural Pollutants

No agricultural pollutant poses a severe national environmental threat, although several are locally or regionally severe (Table 7.1). Agricultural biocide pollution was sharply curtailed by regulations introduced before the 1980s. In some cases, such as grain straw shorteners and some herbicides, there is no clearcut evidence that a hazard exists. Pollutants originating outside agriculture—even outside Sweden—are not explicitly analyzed here. These include radioactive soil contamination from the Chernobyl disaster, acid deposition, and increased ultraviolet radiation due to stratospheric ozone depletion.[3]

Nitrogen Leaching and Water Quality

Nitrogen leaching into ground and surface water is widely held to be the most serious pollution problem caused by Swedish agriculture. Among the most dramatic symptoms of eutrophication are episodic toxic algae blooms, fish kills, and a "dead" sea bottom in nearly one-

[3] Soil acidification is also caused by the use of acid-reaction fertilizers and removal of base elements in harvested crops. Neutralizing all sources of acidification would require an average annual application of 170 kg CaO/ha. Since the actual rate has been only half that, the average soil pH is declining. This reinforces doubts about the long-run sustainability of present farming methods (Lidén and Andersson 1989:4).

Table 7.1. Swedish agricultural pollution problems

	Problem					
Source	Surface water eutrophication	Groundwater toxicity	Soil toxicity	Gaseous emissions	Workplace safety	Food safety
Animal wastes	T, I[a], L	M, L	—	T, I, M[b]	M[c]	—
N fertilizer	T, I[a], L	L	—	M	—	—
P fertilizer	L	—	I[d]	—	—	M[d]
Cultivation methods (bare land, etc.)	T, I, L	L	—	—	—	—
Sewage sludge	L	L	L	M	—	M
Biocides	—	L	?	M	M	I
Soil erosion	L	—	—	—	—	—

Note: — = no problem or negligible risk, T = adverse transnational environmental impacts, I = intermediate degree of national severity, M = minor national problem, L = locally severe, ? = negative effect uncertain; scientific debates unresolved.

[a] Includes toxic effects of poisonous algae.

[b] Ammonia and nitrogen oxides considered to be of intermediate severity; methane and CO_2 of minor severity.

[c] Health effects of stall gases

[d] Due to cadmium contamination.

third of the Baltic. The Baltic Sea is the world's largest estuary and is very sensitive to nutrient loading because of its slow water circulation. Several countries contribute to Baltic nitrogen pollution, and agriculture accounts for 40 percent of Sweden's (anthropogenic) share (NEPB 1990a).

On average, Swedish farms leach 17 kg/ha/year of nitrogen into ground and surface water, a small magnitude compared to neighboring countries such as Germany (29 kg/ha) and Denmark (91 kg/ha). Nationwide, farming generates about 30 percent of total nitrogen leaching from 8 percent of Sweden's land area. Leaching in intensively farmed, sandy soil regions of southern Sweden, however, typically exceeds 50 kg/ha, and agriculture's share of the total nitrogen load reaches 80 percent. Further, about one hundred thousand southern residents depend on groundwater sources that exceed the national health standard of 50 mg nitrate/l (Haxsen 1990, Lidén and Andersson 1989).

The national goal, set out in a 1985 parliamentary action plan, is to reduce agricultural nitrogen leaching to 50 percent of the 1987 level by the year 2000. Although serious leaching is distinctly local and regional, a 20 percent cut in aggregate application of soluble nitrogen was set as an operational target to achieve the 50 percent reduction in leaching. (Recent agreements among the north European governments would require a 50% cutback in Swedish nitrogen effluent to the Baltic and North seas by 1995, but it is not clear how much would come from agriculture.)

The package of policy measures includes regulations, subsidies, targeted research, and extension services. Acreage set-asides aimed at restricting crop production also made a minor contribution in the latter 1980s. Studies show that the economic cost of reduced leaching varies from a few kronor to several hundred kronor per kilogram, depending on the means employed. Thus the selection and combination of policy instruments matter greatly (Andréasson-Gren 1991; Johnsson 1990; MOE 1990a; NBA 1990a, b).

In Sweden as in other nations, attempts to design an effective national policy are limited by four basic facts. First, agricultural leaching is largely a nonpoint process, so effluent from individual fields and farms cannot be monitored directly. Second, there are time lags, first between farm operations and leaching and then between leaching and environmental degradation. This makes it difficult to assess and control pollution stemming from specific cultivation methods. Third, leaching varies greatly from site to site and year to year, depending on aquifer geology, soil parameters, climatic conditions, crop mix, and cultivation practices. Finally, environmental damages depend on the vulnerability of particular aquatic ecosystems. It is estimated that only 10 to 25 percent of nitrogen applications cause serious damage. One can easily imagine the high administrative cost of fine-tuning a regulatory system to capture all these sources of variability.

Only a few partial attempts have been made to estimate the social cost of nitrogen leaching. One analysis of eutrophication in coastal waters provides a rough notion of two economic opportunity costs: lost recreational fishing (using contingent valuation methods) and lost profits in commercial fishing and aquaculture (using a simulation technique). Conservative estimates of these costs are US$220 million and $53 million per year, respectively. On the basis of agriculture's share of the nitrogen load, this analysis leads to a lower-bound estimate of $1.40 per kilogram as the average social cost of leached nitrogen (this cost can be compared with the price farmers paid for fertilizer in 1990: $1.20/kg of nitrogen). Lost fishing activity is, of course, only a fraction of the full social cost of eutrophied coastal and freshwater ecosystems, but no comprehensive measure of current use values, much less of the resource's option and existence values, has been attempted. The narrowness of cost estimates and the wide confidence intervals associated with them mean that the true marginal and average social value of damages is unknown (Silvander 1991, Silvander and Drake 1991).[4]

Fertilizer taxes. An early response to green demands, in 1984, was a 5 percent tax on the nitrogen content of fertilizers. Its primary purpose was to finance research and extension; secondarily, it was viewed as a low-cost way to induce less intensive fertilizer use. The tax was doubled to 10 percent in 1988 as a minor component of the National Action Plan on Pollution of the Seas. Tax revenues (US$20 million in 1988/89) have financed research on improved nutrient management, soil testing, and extension advice on optimum fertilization, all of which are cost-effective measures. More dubious is the recent use of some tax revenues to construct manure storage facilities (see below).

In 1982 a levy on nitrogen was introduced to help finance surplus grain exports. From 1988, the overall tax-cum-levy was 25 percent.[5] Making polluters pay and discouraging fertilizer use were not its pri-

[4] The same contingent valuation studies led to a point estimate of $340 million for preserving groundwater quality, but this figure is not statistically significant (see also Hanley 1989).

[5] The levy was further increased in 1991 as part of agricultural deregulation.

mary goals, but input taxation does contribute to those objectives. Studies of demand responsiveness indicate that the 25 percent surcharge reduced aggregate nitrogen application between 5 and 12.5 percent in the latter 1980s.[6] Since the linkage between fertilization and leaching rates is tenuous, the cutback in pollution was probably much smaller. It is noteworthy that farmers appear to have cut their fertilizer use by more than the economic optimum. Though the higher input price and improved extension advice induced some farmers to eliminate unprofitable applications, others probably responded to the implicit moral condemnation of their role in fish kills and dying sea bottoms (Johnsson 1990).

There has been a running controversy about the goal attainment and cost-effectiveness of nitrogen taxes. Skeptics stress that nonpoint source pollutants cannot be taxed effectively because of the practical impossibility of monitoring. A proxy tax on an input, instead of one on the output of pollutants, is also deficient, since equivalent amounts of fertilizer have radically different environmental impacts depending on soils, climate, and cropping conditions. Fertilization at 100 kilograms of nitrogen per hectare can cause anywhere from 2 to 80 kg of leachate; therefore, a uniform tax will not be cost-effective. Specifically, Sweden's uniform tax does not effectively target the southern plains, where fertilizer-intensive production, sandy soils, and severe eutrophication and nitrate poisoning impose high social costs. Fertilizer cutbacks in areas where leaching poses little problem depress crop yields and farm incomes with little corresponding social benefit. In effect, the tax is far too low on sandy soils planted to annual crops in the south and far too high on clay soils planted to grass ley in the north. Theoretically, a regionally adjusted nitrogen tax could remedy negative allocational effects, but there is no clear idea how to prevent black market shipments from low- to high-tax regions.

The direct financial burden of the tax and levy was US$64 million in 1988/89, roughly 1.5 percent of gross farm receipts. The distribution of costs corresponds poorly with the Polluter Pays Principle. A fertilizer tax burdens the grain and oilseed producer more than a live-

[6] The best estimates of short-run demand elasticity for nitrogen fertilizer range from −0.2 to −0.5. Long-run elasticity is probably significantly higher because farmers can make multiple managerial adaptations. On the other hand, the 15 percent levy had a slight adverse feedback effect: by financing the export of surplus grain, it sustains high grain prices, which induce high fertilization rates. There are no empirical estimates of this effect (Drake 1990b, Johansson 1984).

Figure 7.1. Relationship between nitrogen applications, crop yields, and nutrient leaching

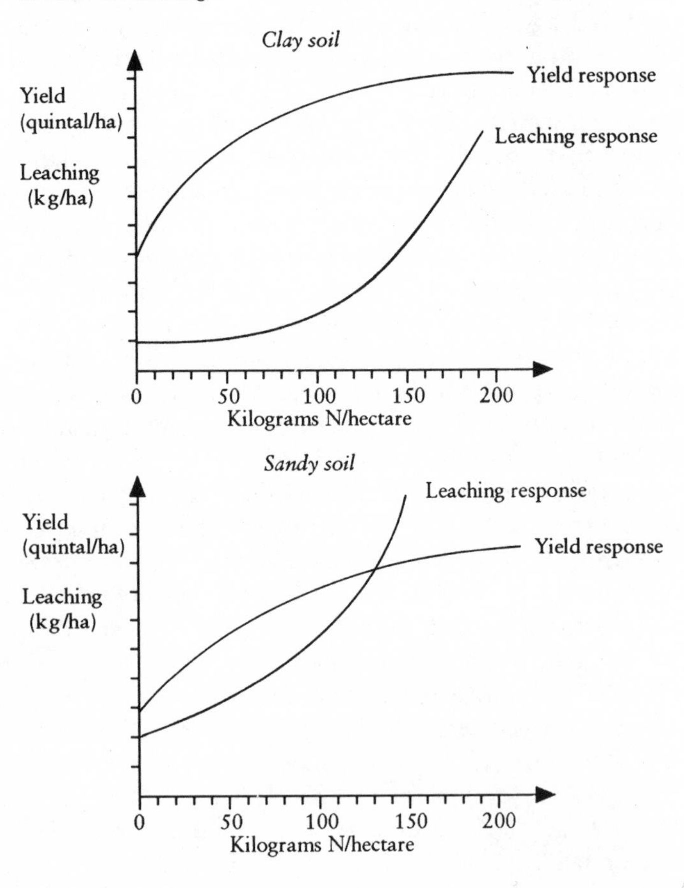

stock producer with the same volume of leachate from animal wastes.[7] A grain/hog producer in a sandy coastal area where leaching is severe pays the same tax as a pure grain producer on inland clay soils who generates far less effluent per ton of fertilizer. This combination of inequity and inefficiency has aroused relatively little political attention.[8] In any case, import restrictions and low demand elasticity for food have shifted most of the cost from farmers to consumers. Without border protection, the burden would fall more heavily on farmers and the fertilizer industry.

[7] Farms that substitute nitrogen-fixing legumes for soluble fertilizer also contribute to leaching but are not taxed.

[8] In principle, the equity problem could be reduced by rebating tax revenues back to producers in some allocationally neutral way, such as a uniform payment per hectare.

Figure 7.1 presents schematic yield response and leaching response curves for clay and sandy soils. At current nitrogen application rates for grain crops—typically in excess of 100 kg/ha—crop yield is little affected by small increases or decreases in fertilizer stimulated by the levy and tax. This is reflected in the flatness of the yield response curves to the right of 100 kg/ha. Proponents of the moderate tax could argue that the steepness of the leaching response curves beyond 100 kg/ha of nitrogen made the tax a fairly cost-effective way to reduce leaching (despite the inefficiencies described above). At higher levels of tax, however, the deeper cuts in fertilizer use would have a higher opportunity cost in lower yields for each kilogram of reduced leaching, especially in clay soil areas (Johansson 1984).

In sum, in a context of excess grain production, it is conceivable that the 25 percent tax/levy was cost-effective, since the social opportunity cost of any loss of grain output was very small. Indeed, the net economic cost of the fertilizer tax may have been negative.[9]

A tax on mineral fertilizer has beneficial indirect effects on nutrient management, although quantitatively important effects would probably require a tax considerable higher than 25 percent. Given time to adjust—or facing a tax above some (unknown) threshold level—farmers would find several measures more profitable: splitting fertilizer applications over the growing season to cut nutrient waste, planting more nitrogen-fixing legumes in ley or as green manure in rotations, and more effective conservation of nitrogen from animal wastes. These indirect effects would further reduce leaching and gaseous emissions. Their likely magnitude in response to different tax rates has not been investigated. If a higher nitrogen price were backed up by complementary research and extension efforts, the impact on leaching could be substantial.

Economists are increasingly interested in marketable fertilizer application permits as an alternative to the tax. (Politicians have not yet shown a similar degree of interest.) If illegal interregional fertilizer shipments could be prevented, this measure would allow fairly precise regional standard setting, based on the severity of water quality degradation. In bidding on permits, farm operators would have an incentive to minimize their opportunity costs (lower yields and revenues), so the

[9]This analysis implies that the current tax would be less cost-effective if domestic grain prices were reduced toward the world market level and excess production were eliminated. This is relevant to the current debate over whether to raise the fertilizer tax as part of the overall agricultural reform (Drake 1990a, b; Hasund 1990a; Johnsson 1990; Jonasson 1991b.

presumption is that pollution-abatement goals could be achieved at a low real resource cost (MOE 1990a).

Manure Management. New manure management requirements, most with multiyear phase-in periods, are a core element of the nitrogen-abatement package. Measures include restrictions on winter spreading, requirements for storage capacity and earthing-over, financial subsidies, and extension efforts. Public expenditures are to be financed from the fertilizer and pesticide taxes, but most of the cost will be passed on to consumers as long as import protection remains in place.

New regulations were introduced in 1988 and took effect in 1989. Most important is a nationwide ban on spreading from December through February, with an exception granted for manure earthed-over the same day it is spread. Beginning in 1995, August–November spreading in southern and coastal areas will be permitted only if manure is applied to an established sod crop or precedes a fall cover crop. The main costs associated with seasonal restrictions are the need for increased manure storage capacity on a few farms, higher production costs when spreading is shifted from periods of slack to peak labor demand, and yield losses owing to bottlenecks that prevent timely operations such as spring sowing. On the whole, though, appropriate seasonal limitations on spreading, combined with the earthing-over loophole, appear to be the most economically efficient ways to reduce leaching (Brundin and Rodhe 1990, Lidén and Andersson 1989). The regional aspect of the timing restriction also improves goal attainment and cost-effectiveness. In the big picture, however, most livestock farms already have adequate storage capacity and most manure is already spread during the permitted periods, so the aggregate costs and benefits may be small. The main potential problem is that small, economically marginal dairy farms, which contribute significantly to landscape values, will shut down because expected revenues will not cover the additional capital cost.

Beginning in 1989, the minimum storage capacity requirement in south-coastal Sweden is eight months of animal wastes for farms with cattle and ten months for hog and poultry operations (compared to the previous six months). This requirement is backed by a 20 percent subsidy on investment in expanded storage facilities, up to a maximum of US$4,000. The requirement is fundamentally flawed. Since it is normally possible to spread manure in an environmentally acceptable way in southern Sweden for much longer than the two to four months per year implied by the capacity requirement, the marginal social benefit will be small. Indeed, on many farms livestock actually graze longer than four months, producing very little stall manure. On the positive side, greater storage capacity will allow better nutrient conservation and utilization, will reduce air as well as water pollution, and could reduce local odor and sanitation problems. (No quantitative estimates of these effects were made before the policy decision.)

Despite these grounds for skepticism, an advisory panel has recommended that storage requirements be extended to other regions. According to its calculations, nationwide compliance would reduce nitrogen leaching by about 3 percent (1400 metric tons) at an incremental cost of at least $40 per kilogram (MOE 1990a, NBA 1990a). This is far higher than the incremental cost of many alternative measures, and it is difficult to imagine that the incremental social benefits could justify costs of this magnitude. As a negative side effect, some economically marginal livestock operations would be shut down because they could not recover the capital outlay.[10] Given the existing overproduction of pork and dairy products, this may have made sense in narrow economic terms, but its adverse regional development and landscape effects seem to have been discounted (Hasund 1991).

The economic costs and environmental impacts of alternative manure handling technologies have not been fully analyzed, but the available studies suggest that a cost-effective reduction of leaching from animal wastes depends far more on the details of technical and management systems than on physical storage capacity. For example, liquid manure systems cut nitrogen loss by about 16 percent more than do solid waste systems; leaching is very sensitive to how and when manure is incorporated into the soil; and inexpensive lids on open storage facilities can conserve nutrients while reducing gaseous emissions by 75 percent (Brundin and Rodhe 1990, Hasund 1990b, Lundin 1988).

Livestock Density. Beginning in 1995, farms will be limited to 1.6 milk cows, 10.5 fattening hogs, or 100 laying hens per hectare. The limits are derived from estimates of farmland's capacity to take up phosphorous from manure over the rotation. Farmers expanding or adjusting their operations before 1995 must meet the requirement immediately. It can be met either by spreading manure on one's own land or by making a long-term contract to spread it on other land that falls below the maximum application rate (NBA 1990e).

[10] In Denmark, Alex Dubgaard estimates that as many as one-fourth of livestock producers would terminate operations if they faced an analogous investment cost (1990b:25).

Most livestock farms already satisfy the stocking limit, but a few hundred intensive livestock operations in the south will probably have to reduce stocking densities, since there is not enough nearby land to absorb the present volume of manure (Lidén and Andersson 1989). The stocking limit will have little nationwide effect, but since the adverse effects of leaching and air pollution tend to be locally concentrated, there could be a major impact in hard-hit areas. The economic costs will not be clear until the noncomplying farmers reveal their responses: they may truck wastes to areas with lower stocking densities, reduce the size of confinement operations, or revert to less feed-intensive production methods. According to Andréasson-Gren (1991), the direct plus opportunity costs of using stocking reductions to cut nitrogen loads to the severely eutrophied Laholm Bay would vary from less than US$10 to more than $300 per kilogram of nitrogen, making this a highly questionable measure.

Green Land. Winter cover crops reduce leaching by absorbing soluble nutrients in the fall and early spring. They include perennial sod crops, fall-planted grain and oilseeds, and aftersown or undersown catch crops that take up soil nitrogen after the main crop harvest. Since the severity of leaching varies so much with soil type, watershed conditions, and local climate, the National Board of Agriculture was assigned to formulate locally differentiated standards. The resulting 1995 targets call for the proportion of autumn-covered land to increase from roughly 40 percent today to 60 percent on farms in southern coastal counties and to 50 percent in the rest of south and central Sweden.[11] Many farms with pastures and extensive ley already meet these standards.

The proposal for mandatory green cover builds on experience with cross-compliance requirements in the grain acreage set-aside scheme. Beginning in 1988, new plantings of cover crops, ley, and softwood forest on former grain land became eligible for set-aside payments. Starting in 1989, set-aside participants in southern Sweden were required to maintain at least 60 percent green land year round (NBA 1990c). These stipulations broadened farmers' land-use options while reducing leaching. Environmental improvement was achieved with no additional expenditure for the set-aside program (apart from a small subsidy to some southern farmers who participated in catch crop ex-

[11] Accepted crops are ley (sod), winter grains, winter oilseeds, late-harvested root crops, perennial fruits and berries, catch crops, and salix, a fast-growing bioenergy feedstock.

periments).[12] If we take the irrational farm price policy and excess production as given, these side conditions in the set-aside program were quite cost-effective.

The NBA calculated that if winter cover reached the 1995 targets of 40 to 60 percent, agricultural leaching would be reduced by 7 percent (3,300 metric tons) at an average abatement cost between US$2.80 and $4.20 per kilogram of nitrogen; this is a whole order of magnitude less than abatement via new manure storage capacity (NBA 1990a, b). The Laholm Bay study reinforces the conclusion that cover crops are an inexpensive way to reduce nitrogen loads. The predicted marginal cost varies from $2 to $9 per kilogram of nitrogen, depending on the necessary cultivation operations. If, however, the cover crop requirement were raised to 80 percent, the marginal cost would rise sharply. In an experiment with three different catch crop systems, the net cost of abatement ranged from $0.50 to $15.00/kg, depending on the cost of establishing cover crops, their economic value, and their effect on leaching under different conditions (Sterner 1990). Because the marginal cost of complying with the regulation and cutting pollution varies so drastically, the Ministry of Environment (1990a) concluded that an economic instrument, such as a tax on uncovered land, would be a more flexible and efficient way to achieve abatement targets than the quantitative regulations actually adopted by the Riksdag. A tax creates an incentive for farmers to seek the least-cost method of compliance.

Nonetheless, the evidence suggests that increasing ley and fall-sown land is among the least expensive ways to cut leaching, perhaps by 7 percent. Much of the cost will be borne by consumers as long as border protection remains in place. Within the farm sector, producers specializing in oilseeds and summer grain will be more burdened than farms whose acreage is already in the permitted crops. The problem is greatest for producers on heavy clay soils, who may face costly timing bottlenecks in their fall operations. The Polluter Pays Principle is partially subverted, since these farms typically add little to pollution.

Enforcement could be a stumbling block. Although policing the regulations is likely to be difficult and expensive, it was not even discussed in the report that paved the way for the policy decision, and penalties for noncompliance have not yet been set. If penalties are small, farms with high compliance costs will be tempted to cheat, and

[12] The participating farmers had to bear a cost to receive set-aside payments. In an experiment monitored by Sterner (1990), the net cost of growing catch crops ranged from US$6.50 to $120 per hectare (after deducting the value of crops for fodder or as a nitrogen source).

enforcement will require many field inspections or the use of remote-sensing technology, which could provoke hostile farmer reactions to the Big Brother state. Stiff penalties would have a greater deterrent effect and necessitate fewer on-site checks, but they would probably offend farmers' sense of fairness, since the offense is perceived to be minor and climatic conditions can make compliance difficult in some years. Historically, state authorities in agriculture as well as other sectors have been able to rely largely on moral suasion to secure a high degree of voluntary compliance.

If the cited estimates from experimental and simulation studies are correct, expansion of cover crops is among the most cost-effective means of cutting leaching; indeed, the 1995 acreage targets are probably too modest. An even larger percentage of green cover, substituting for less cost-effective measures, would lower the social cost of reducing leaching 50 percent by the end of the century.

Sensitive Areas and Wetlands. Since 1984, the National Environmental Protection Board has been empowered to designate environmentally sensitive areas where especially strict chemical and manure management practices are enforced. Up to 1990, the regulation was implemented only in two intensely farmed southern areas with severe leaching problems: the watersheds of Lake Ringsjön and Laholm Bay. Laholm Bay has come to symbolize fish kills and dead sea bottoms in Swedish popular consciousness. Exceptional regulations include accelerated compliance deadlines and stricter requirements for animal density, manure storage capacity, earthing-over manure, soil testing, and mandatory farm nutrient management plans. Livestock producers have been the focus of extension efforts to educate producers about their duty and their economic interest regarding nutrient conservation and reduced use of mineral fertilizer on manured fields. This voluntary program has had considerable success and is quite cost-effective. The key fact is that farmers, facing a cost-price squeeze, have been eager to try resource conservation methods that might improve their bottom line. As a result of these voluntary and mandatory measures, agricultural nitrogen leaching on the Swedish West Coast is reported to have been cut by half in the 1980s (Andersson 1985, Andréasson-Gren 1991, Kumm 1984, Sandberg 1990).

In the latter 1980s, subsidized reconversion of arable land to wetlands was one of the green measures prompted by excess production. Where aquatic ecosystems are vulnerable to eutrophication, a fringe

Table 7.2. Official measures to reduce nitrogen leaching by 50% between 1990 and 1995

Measure	Reduction in leaching	
	Metric tons	Percentage of 1987 leaching
Limits on animal density Manure management plans	2,300	4.7
Manure storage capacity Spreading restrictions	5,400	11.1
Research (spreading techniques) Certification of equipment	2,400	4.9
Green land	3,300	6.8
Reduced fertilizer use (tax, levy, lower crop prices)	5,000	10.2
Reduced grain acreage (500,000 ha)	4,000	8.2
New protein fodders	1,000	2.0
Fewer cows	1,000	2.0
TOTAL	24,400	49.9

Source: NBA 1991a.

benefit of wetlands is their capacity to clean nutrients from watercourses by binding them in organic matter.

Table 7.2 summarizes the expected contributions of various nitrogen-abatement methods listed in the government's 1990 proposal to cut leaching by 50 percent by 1995 (GP 1990c). The table describes a comprehensive policy, but one that is not well crafted to minimize costs. Indeed, little is known about the direct and opportunity costs of most instruments, and the economic case for several of them is weak. For example, the government estimates that to reduce leaching via increased manure storage capacity costs US$46 per kilogram on average, versus $3 to $4 per kilogram via expansion of green land. Existing knowledge does not allow us to project how much further the most cost-effective instruments, such as green cover and manure spreading requirements, could be increased before they encountered physical limits or diminishing economic returns.

Table 7.3 summarizes measures to protect Laholm Bay, which have been subjected to a more thorough analysis. The estimated marginal costs reflect the special conditions of this region, where soils are sandy and manure and fertilizer applications have been heavy. Some measures seem clearly superior to others, at least at the abatement levels studied. For example, the highest cost via cover crops is about the same as the lowest cost via lower livestock density. The wide range of possible costs for a single instrument, such as manure spreading

Table 7.3. Marginal costs of nitrogen reduction in Laholm Bay

Measure	Marginal cost (US$ / kg N)	Aggregate effect (metric tons of N)[a]
Agricultural		
Fertilizer reductions	2.3–58	798
Timing of manure spreading	2.5–38	190
Cover crops	2.0–9	75
Establishing grass ley	6.3	109
Reduced livestock density	8.5–326	150
Planting energy forest	4.3	300
Establishing wetlands	0.7	430
Nonagricultural		
Reduced sewage discharge	8.3	600
Reduced vehicle emissions via speed limits	32.3–269	325

Source: Andréasson-Gren 1991:11.
[a]Upper-bound estimates of nitrogen abatement.

requirements, underlines the exceptional difficulty of fine-tuning policy instruments to minimize the total cost. For nitrogen reduction generally, we find it surprising that there has not been a greater political will to make more extensive use of the most cost-effective measures.

Air Pollutants

Agriculture contributes to gaseous emissions mainly via denitrification and evaporation of animal wastes. Their precipitants combine with other pollutants to cause acidification and eutrophication. They can also create workplace hazards (stall gases) and off-farm odors, though strict standards for ventilation of livestock confinement facilities have minimized the workplace safety problem. Emissions add in a minor way to international acidification, eutrophication, ozone depletion, and greenhouse gas accumulation. (Sweden's complicity is small compared with that of neighboring Denmark and the Netherlands; see Chapter 10.)

Nitrogen compounds—ammonia, laughing gas (N_2O), and nitric oxides—contribute most to air pollution; methane from ruminants and carbon dioxide from fossil fuel combustion are the main greenhouse gases. The farm sector accounts for about one-third of Swedish total emissions of nitrogen gases. Most important among these is ammonia, which in some locales is a major contributor to soil acidification, eutrophication, and even mortality of some broadleaf and coniferous species.

Virtually no measures are explicitly targeted against agricultural air pollution, since it is not viewed as a high-priority problem. Even so, the nitrogen fertilizer tax, ban on winter manure spreading, and decree on prompt earthing-over of manure reduce nitrogenous emissions substantially. In the case of ammonia, agriculture is responsible for 90 percent of Sweden's annual emission of 50,000 to 70,000 metric tons. Evaporation of livestock wastes in stables, storage facilities, and spreading operations is by far the largest source.[13] As much as 50 percent of the nitrogen in manure and urine is wasted and becomes an environmental nuisance.

The government's 1990 environmental bill for the first time established a target for ammonia abatement: a 25 percent reduction in emissions by 1995. Exploration of the implications of a 50 percent reduction by the year 2000 was also initiated. The National Board of Agriculture was charged to formulate an implementation plan for the livestock-intensive southern and southwestern regions within one year. Planning for the remainder of the country has not begun. State officials apparently hoped that contraction of dairy herds after the 1990 food policy would do the job without further discretionary measures.

Phosphate Runoff

In the 1960s and 1970s, phosphate eutrophication of freshwater bodies was a prime focus of Swedish environmental policy, but outside a few intensive production areas, agriculture was not a major contributing cause. Better sewage treatment has improved water quality and correspondingly reduced concern about phosphorus. The 1988 environmental legislation simply states the objective of "reducing phosphorus effluents in regions that are seriously affected." The previously cited international accord to cut nutrient emissions to the Baltic and North seas in half by 1995 includes phosphorus, so further national action, focused on nonagricultural sources, is forthcoming.

At present, the annual phosphorus runoff from arable land averages less than 250 grams per hectare (620 metric tons in toto), and one-fourth of this would occur even without any soil runoff (80 to 190 tons of phosphate detergents from cleaning operations on livestock farms end up in water bodies) (Olsson and Löfgren 1989). In addition to effects on local water quality, Swedish phosphate use contributes

[13] Smaller losses also occur during the growing season, especially in heavily fertilized grain crops that face climatic stress when plants are setting grain.

about 0.5 percent to annual worldwide depletion of the finite stock of rock phosphate.

Although there is no target for cutting agricultural phosphorus pollution, it was in fact sharply curtailed when phosphate fertilizer sales declined from 53,000 to 30,000 metric tons per year over the 1980s. Manure spreading restrictions will curb phosphate runoff further (SS 1991a). Since the agricultural phosphorus problem has low status, little analysis has been done on alternative ways to reduce runoff. More attention is paid to superphosphate's cadmium residues, discussed below.

As with nitrogen, a combined tax and levy (currently 30%) is imposed on phosphate fertilizer. Most of the revenues finance grain exports, with the remainder supporting research on reduced-chemical production and environmental protection. The fees had a minor role in reducing superphosphate use in the 1980s. A Norwegian analysis draws ambiguous conclusions about the cost-effectiveness of a phosphorus tax to limit runoff. At Sweden's low application rates, the opportunity cost of lower yields may well outweigh the social benefit from incrementally lower pollution (Johnsen 1990).

Cadmium Contamination

Annual cadmium deposition averages 1 gram per hectare of farmland, causing some concern about its toxic effects in food and its concentration in the soil. About 45 percent of cadmium deposition stems from phosphorus fertilizers, 36 percent is due to precipitation, and 4.4 percent comes from sewage sludge applied to farmland. The share from sludge is much higher on land that has been enrolled in sewage recycling, yet apart from a small number of local test sites, crop sample analyses have found cadmium residues well below the level deemed harmful to humans (Lidén and Andersson 1989, SS 1990b).

An extensive program to eliminate cadmium from other economic sectors and purge it from the municipal waste stream will further reduce soil contamination from both precipitation and sludge applications. The policy bans cadmium from several uses, taxes cadmium batteries, and imposes tough standards for "scrubbing" fumes. A ceiling of 4 grams per metric ton of dry matter has been set on cadmium in sludge applied to cropland (this will be cut to 2 grams in 1995). Sludge applications are limited to one ton of dry matter per hectare every five years, and spreading on vegetable crops is prohibited.

Spreading is not permitted within ten meters of an open water body. Taken together, these requirements are considerably stricter than those in other European countries (NEPB 1989, 1990b).

To avoid restrictions on cadmium in fertilizer, the dominant superphosphate producer reached a voluntary compliance agreement with the state in the early 1980s, committing itself to import raw materials with the lowest cadmium content and to publish an annual declaration of the quantity of cadmium in its product. Swedish researchers have also developed a method for "cleaning" phosphate rock. The result has been a reduction in cadmium deposited in fertilizer from 5 grams per hectare in 1970 to 1 gram in 1989. Based on the available evidence, there is no clear biomedical case for restricting cadmium to such a low level (Lidén and Andersson 1989, MOE 1990a). Indeed, we contend that an appropriate tax on all sources of cadmium could achieve the politically determined level of cropland deposition at a lower opportunity cost.

The hot political issue at the end of the 1980s was LRF's categorical refusal to allow sewage sludge on arable land. Among the boycott's effects were increased sludge burning and greenhouse gas emissions, rising marginal costs for landfilling sludge, and wastage of recyclable nutrients and organic matter (see section 7.3). The added cost to municipalities was millions of dollars each year (Hahn 1991).

Agricultural Biocides

Objections to agricultural uses of chemical insecticides, fungicides, and herbicides have centered on health risks to consumers of food and groundwater, safety hazards to farm workers, and ecological disturbances, on and off the farm. According to Lidén and Andersson:

> Residues of pesticides, mostly phenoxy acids, have recently been found in water supplies in Sweden. Inaccurate handling is suspected to be the main reason but leaching of pesticide residues from treated soils is another possibility which cannot yet be excluded. Pesticide pollution of drinking water is not yet considered to be a health hazard, but according to policy guidelines no pesticide residues should be accepted in groundwater. . . . [In addition] ditch banks and other adjacent biotopes are often exposed to pesticides when crops are treated, either as a consequence of wind drift or as a consequence of active actions. (1989:5–6)

The health hazard from biocide residues in Swedish food has been judged very small. Hazards to farm workers are also small but not insignificant. Populations of species sensitive to biocides are drastically reduced on and near treated fields, but chemicals currently threaten no species with extinction. Populations of birds such as pheasants have also been reduced by the loss of food (and in past times by accumulation of toxic compounds in the food chain). Finally, future economic losses could be caused by the growing resistance of pests and pathogens to biocides (Fogelfors et al. 1992).

Since 1984, a mix of scientific evidence, media alarms, and pressure from environmental organizations has led to a series of loosely coordinated policy initiatives centering on testing and licensing of biocides, research, farmer training and farm-level handling regulations, certification of spraying equipment, and taxes on agricultural chemicals. The core objective has been to promote pest control methods with lower health and ecological risks. To this end, an action plan was legislated in 1985, stipulating that all agricultural chemicals be relicensed every five years under tightened standards. The National Chemicals Inspectorate has instituted a rigorous procedure of risk-benefit analysis and sponsored research on chemical and nonchemical substitutes with lower risk profiles to replace currently licensed biocides. By these means, many of the most hazardous products have been eliminated. As economists, we are skeptical, however, of absolute rulings that totally exclude substances that pose no significant hazard if used in small doses and with proper handling. Certification procedures are not well designed to balance health and environmental objectives against the economic benefits of crop protection and higher yield.

As in other industrial nations, the state sponsors research and testing of a wide range of potential technical innovations, including less toxic and less persistent chemical compounds, ultralow-volume application equipment, pest and pathogen monitoring and warning systems, biological and botanical control methods, and genetic resistance. Monitoring of pesticide residues in water, plant tissue, and food is another area of growing public investment. Five plant-protection centers now train extension officers in integrated pest management.

Another tack is to improve handling of pesticides, since mishandling is the principal cause of workplace hazards and environmental damage. As of 1990, applicators must take a three-day course on all aspects of pesticide management and application and pass a test to become certified. Every five years they must take a refresher course for recertification. Spray equipment on farms is periodically tested for its dosage efficiency and safety. Equipment testing is combined with extension advice to farmers (NBA 1990d). In section 7.3 we describe a separate program that subsidizes farmers who convert to completely chemical-free production.

An early initiative was the introduction of a pesticide tax in 1984. The initial rate, US$0.70 per kilogram of active substance, was doubled in 1988 along with the fertilizer tax. As with fertilizer taxes, revenues are channeled primarily into research. As an incidental effect, pesticide use is reduced by the tax, but this effect is extremely variable. Since the tax is based on the weight of active ingredient, new high-potency, low-dosage chemicals are effectively tax-free. And since the economic and ecological effects of a flat tax per kilogram are so variable, the tax is not an effective incentive[14] (NBA 1989a). (Newly proposed legislation, discussed in Chapter 9, corrects this defect by setting a tax per dose, based on each chemical's recommended dosage per hectare and its environmental risk.) As with fertilizer taxes, the social stigma against biocides may have a moral or psychological effect, since farmers have cut pesticide use by considerably more than the higher price would justify on narrow profit-and-loss grounds (Sundell 1982).

The 1987 Action Program on Agricultural Chemicals set a national target of cutting the 1981–85 tonnage of active ingredients in half by 1990. An even more ambitious resolution, to cut usage 75 percent by 1995, was voted by the Riksdag in 1989. Subsidiary goals include reductions in the treated acreage and conversion to less dangerous biocides.

The use of many chemicals was already below the economic optimum in 1987, and one analysis indicates that the further 50 percent reduction—with the limited technological alternatives existing at present—would have a social opportunity cost of US$50–65 million. The proposed 75 percent cutback would raise that cost to at least $70 million and possibly as much as $430 million (Fogelfors et al. 1992). It should be kept in mind that limiting the use of chemical pest and weed controls can have negative near-term environmental side effects, especially increases in fossil fuel consumption and soil compaction when mechanized weed control replaces herbicide treatment.

Figure 7.2 shows that there was indeed a sharp cutback in the volume of all types of agricultural chemicals after the peak year of 1986, although deciphering the causes is problematic. Along with a higher

[14] The cost of chemicals per treated hectare ranges from $0.10 to $22.00, depending on the compound and application method; thus the tax ranges from less than 1 percent of cost to over 50 percent, but not in any systematic relationship to a substance's environmental risk (NBA 1989a).

Figure 7.2. Sales of biocides in commercial agriculture and horticulture, 1981–1990

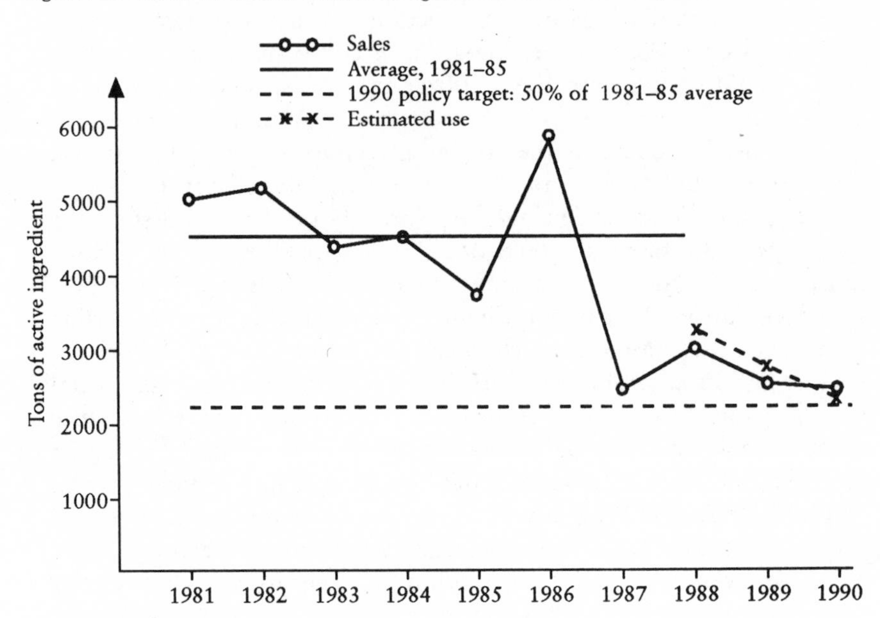

Source: NBA 1991b.

chemical tax, there were also less severe insect and fungus problems, lower net grain prices, less grain acreage, more potent chemicals, and new low-volume spraying equipment. Note too that the high year-to-year variance makes identifying trends problematic and that only insecticides have been sharply curtailed relative to crop acreage. The aggregate reduction from 1982–85 to 1989 resulted from cuts of roughly one-fourth in both hectare applications and the average dose per application (Table 7.4).

7.2 Promoting Farmland Preservation and Biological Diversity

In economic jargon, agriculture is a multiproduct industry and farmland is a multiple-use resource. Beyond its contribution to present consumption and to future food security, the agrarian landscape has

Table 7.4. Reduction in biocide applications

Year	Proportion of total crop acreage treated with biocides (percentage)		
	Herbicide	Fungicide	Insecticide
1988	50	8	30
1989	50	9	12
1990	48	8	11

Year	Applications and dosages	
	Hectare applications (millions)	Average dose (kg/ha)
1982–85 avg.	4.7	0.9
1986	7.5	0.7
1987	3.3	0.7
1988	3.5	0.8
1989	3.5	0.7

Sources: NBA 1991b, SS 1991.

cultural-historic, recreational, aesthetic, and ecological significance. Pieces of farmland of course differ greatly in their capacity to supply these public goods, and that fact underlies farmland preservation policy.

General farm price policy, plus special supports for north Sweden, rationalization measures, and tax provisions, have all helped to maintain a positive economic rent on land that would otherwise be economically submarginal. By these means, cultivation and an open landscape have been maintained on some low-productivity fields and in disadvantaged regions. The aggregate effect of these untargeted actions on the landscape has been greater than explicit farmland preservation measures. But apart from the Norrland support, their net influence on the detailed character of the rural landscape has been negative. Price supports and rationalization measures, in particular, have shifted land use away from biologically rich and scenically attractive grazing landscapes toward a more monocultural grain and row-crop landscape (Drake 1989, Hasund 1991).

Five types of targeted measures have been devised:

- Nature reserves
- Contractual payments to owners of biologically valuable land (NOLA)
- Environmental considerations in the Land Care Act, restricting certain rationalization measures

- Reconstruction subsidies for wetlands and landscape elements (NYLA)
- Afforestation assistance.

Nature Reserves

Over the decades, hundreds of nature reserves have been established by legislative mandate. The two hundred that are situated in farm landscapes encompass twenty thousand hectares, including pastures, forest lots, etc. The reserves are legally protected, and private owners are compensated for restrictions on their use of the land. Nature reserves have been important for protecting specific valuable objects, such as a field island with ancient oaks or a flora-rich streambed, but they extend to less than 1 percent of arable land. Establishing new reserves has frequently been politically contentious, since owners and tenants tend to view reserves as unfair intrusions on their private use rights. The state's management is often bureaucratically cumbersome and costly. For these reasons, new reserves are established only where there are very special resources, such as endangered species, that need legislative protection or professional treatment. Otherwise, the policy emphasis has switched to NOLA, voluntary contractual arrangements, described below.

NOLA Contracts

As a supplement to the universal requirements of the Land Care Act, a voluntary program, NOLA (*naturvårdsinsatser i odlingslandskapet,* or nature conservation efforts in the agrarian landscape), was established in 1986 to protect biologically rich meadows and *hagmark,* the special grazing land dear to Swedish hearts.[15] As explained, *hagmark* is actually a gestalt of historical, aesthetic, recreational, and ecological meanings. NOLA uses three- to ten-year land-management contracts, negotiated between landowners and a board of their county governments. A management agreement for pasture land, for example, includes stipulations on minimum grazing intensity and no fertilizer use. In effect, owners are paid a rent for the public goods they provide.

NOLA is financed by the state budget, but in many cases municipalities and regional governments contribute additional funds. The program began with less than US$1 million of state funds in 1986 and

[15] NOLA had a forerunner, the 1960s experimental program called Nature Conservation Measures in Cultivated Areas.

with rental payments typically in the range of $50 to $100 per hectare. The budget was doubled in 1987 and again in 1988, when the eligibility criteria were broadened. By 1990, total enrolled acreage had reached thirty-five thousand hectares, about 7 percent of Sweden's remaining natural pastures and meadowlands (Axelsson 1988).

At one level, the NOLA experience has been quite positive. For a small total expenditure (and a level of compensation per hectare less than the public would be willing to pay), a nontrivial fraction of unique and highly valued farmland is being maintained. In contrast to the case of nature reserves, owner-operators maintain the land without costly state management. By communicating society's high valuation of farmers' activity, the program also has had a positive psychological effect, convincing some farmers to continue economically submarginal operations (Hasund 1991). The rental payments are consistent with PCP, the Producer Compensation Principle: producers of beneficial nonmarketed services are compensated for their contribution to social welfare. During the 1989–90 food policy debate, the NOLA program also had a positive educational effect in helping to clarify that not all farmland is equally worthy of public support. Finally, in the context of overproduction, NOLA's thirty-five thousand hectares added only minimally to the meat supply.

At other levels, however, doubts arise. First, goal attainment would be higher if program administrators were free to select the most valuable pieces of land, instead of selecting only from parcels that farmers choose to enroll. Second, cost-effectiveness would be greater if administrators could target land parcels that, without special support, would be abandoned, afforested, or converted to nonfarm uses. Since most NOLA participants have a strong attachment to their land, it is arguable that many are paid to manage the land more or less as they would without compensation. There is little doubt, however, that over the long run targeted compensation is necessary to maintain the remaining meadow and pasture land: two-thirds of it was lost between 1945 and 1988, in spite of the owners' sentiments. Finally, and most important, if our contingent valuation estimates are anywhere near the mark, the Swedish public would be willing to pay much more than is currently budgeted to protect this prized but economically vulnerable land (Drake 1988).[16]

[16] A cautionary note: negotiation and enforcement of contracts for NOLA land have a fairly high administrative cost, which absorbs part of what the public is willing to pay for farmland protection.

Environmental Consideration Rules

The Land Care Act, originally introduced to protect arable land for its production capacity, was rewritten in 1984 to reflect the growing interest in farmland as a multiple-use resource. Nature conservation was given high priority because three-fourths of Sweden's roughly four hundred threatened plant and animal species are associated with the agricultural landscape.[17] Broad nature conservation statutes require that all citizens consider the impact of their actions on biotopes. In agriculture, this has been supplemented by the Land Care Act's "consideration" paragraph: "When cultivating arable land and making other agricultural use of the land, consideration shall be given to nature preservation." Fairly detailed operational rules and recommendations were set forth in 1984 by the National Board of Agriculture in collaboration with the National Environmental Protection Board. The key restrictions were that field islands of significant scenic, biological, or cultural-historic interest may not be removed; trees and hedgerows planted for wind protection may not be removed; and wetlands that are habitat for endangered or rare species may not be filled or destroyed. There are several additional nonbinding recommendations. In the event, the penalty system does not contain strong incentives for compliance, and enforcement has been somewhat lax. After the publication of Hasund's (1991) critical assessment of the policy's actual effect, the wording of the consideration rules was made tougher and more detailed (Hasund 1991, NBA 1984).

NYLA Reconstruction Subsidies

A different compensation method aims to reverse the long history of landscape rationalization in grain-producing plains regions. The irony that forty years of state subsidies for landscape rationalization are now shifting into reverse is not lost on observers. NYLA (*nya inslag i landskapet*, or new features in the landscape) compensation was introduced in 1989 and budgeted at US$5 million per year for three years. It pays farmers to recreate wetlands, establish hedgerows, plant groves of trees on cultivated land, and create other elements to enhance visual and biological variety. As mentioned, wetland reconstruction may also reduce eutrophication in some coastal areas. These measures were justified by their contribution to reducing grain produc-

[17] In the Swedish regulatory framework, "threatened" species include those that are endangered, vulnerable, rare, or require special care.

tion, and the three-year limited reflected expectations that overproduction would soon be eliminated.

Participation is voluntary and presumably based on the farmers' calculus of costs and returns. Subsidized costs can include capital investment, loss of cultivable land, and higher operating expenses due to less rational field layout. The state reimburses investment outlays up to $3,300 per project; land permanently taken out of price-supported crops receives a one-time compensation of $1,600 to $2,400 per hectare, depending in part on past yields. At these compensation levels, the budget covers merely one to two thousand hectares per year (NBA 1990c).

The small budget means that NYLA will have a minor impact on the overall appearance of the plains region, although it may be strongly positive at certain highly visible sites with destitute landscapes. In any case, recreating landscape elements is very costly per unit of biological or recreational benefit, especially compared with protecting existing elements. Because farm-level projects are so varied, no overall conclusion can be drawn about whether the combined social benefits—greater landscape variety, reduced leaching, and lower grain output—justify the program's costs (Hasund 1991).

With an economically sensible grain-pricing policy, a prime NYLA benefit—reduced surpluses—would be eliminated. But there is no certainty that the ensuing cutback in intensive crop production would stop further aesthetic and ecological deterioration of the landscape. Large-scale spruce or energy forest plantings, for example, would be worse than grain by most noneconomic criteria. Some type of targeted incentive to create particular landscape characteristics might therefore continue to be justified.

Afforestation Assistance

The option of planting trees on arable land in the 1989 and 1990 acreage set-asides has landscape implications far into the future. Because of the high establishment cost and delayed payoff, conifer plantations were entitled to three times and deciduous forest five times the normal yearly set-aside payment. The premium for deciduous trees reflects a popular aesthetic preference to reduce the dominance of spruce and pine in the forestscape. For many years, the LRF had asserted that protecting open farmland was a quasi-sacred national duty. When it became clear that afforestation might be the only way to prevent a severe deflation of land assets with lower grain prices, however, LRF began to appeal for afforestation subsidies.

The aesthetic and habitat effects of conversion depend on the mix of tree species and whether the planting locale was previously dominated by forest, a mix of agriculture and forest, or open fields. Carefully located deciduous or mixed stands in an area with horizon-to-horizon grain production (not very common in Sweden) might actually increase scenic and biological variety. But in general, state support for plantations, especially in already heavily forested regions, can be criticized on grounds that citizens are willing to pay to preserve the remaining open farm landscape.

In terms of ecological impacts, forest planting on heavily fertilized cropland typically reduces annual nitrogen leaching from 20–30 kilograms per hectare to 5, but afforestation subsidies were not targeted to regions where leaching is a serious problem.[18] In recent debates over national strategy to restrict greenhouse gas emissions, attention has also been paid to an expanded forest's increased carbon dioxide uptake and the potential of renewable bioenergy to reduce net carbon emissions. By some estimates, subsidized tree planting would actually be more cost-effective than the carbon emissions tax that was introduced in 1991 (Kumm 1989, MOE 1990a, Sedjo 1989).[19] In general, afforestation has such complex long-term economic and environmental impacts that it is impossible to make any sweeping pronouncement about the cost-effectiveness of subsidizing tree planting as part of a scheme to cut grain production.

7.3 Sustainable Resource Use and Alternative Agriculture

Long-term sustainability was introduced as one component of the environmental objective in the 1985 farm legislation. The buzzword "sustainable agriculture" has many connotations, and a broad farm-level definition could include any management practice that uses fewer

[18] Claims for reduced leaching must be qualified, since studies of the Laholm Bay watershed indicate that much of the nitrogen stored in a growing forest's biomass is released as leachate after clearcutting (Sandberg 1990).

[19] If the uptake of carbon dioxide by growing trees is valued at the rate of the present emissions tax (US$0.05/kg of CO_2), the annual social benefit is about $150/hectare/year for the first several decades in the forest's growth cycle (with variations by species and location). When added to the economic value of timber growth, this nearly doubles the discounted present value of investment in forest (Kumm 1989 and authors' calculations).

nonrenewable inputs, maintains soil biological activity, limits soil compaction and erosion, prevents irreversible conversion of farmland to other uses, or minimizes adverse impacts on surrounding ecosystems. Our primary focus here is the long-run productive capacity of Swedish agricultural land, but we also discuss reduced-chemical farming, waste recycling, and production of bioenergy.

Protecting Arable Land Resources

As mentioned, past measures to preserve farmland derived primarily from the national food security objective. The 1979 Land Care Law updated previous legislation to protect farmland from topsoil mining and prevent afforestation of arable land.[20] In the decade leading up to the 1979 law, the central environmental issue bearing on agriculture was conversion of arable land for urban expansion, roads, airports, and the like. A series of statutes established a highly regulated physical planning system, which, together with high farm prices and incomes, cut farmland conversion from 6,000 hectares/year in 1965–75 to only 2,000 ha/yr in 1975–80. Where feasible, less fertile land was preempted for development (SPA 1983, SS 1982). Legislative action culminated in the 1987 Natural Resources Act, which stated that "arable land may be acquired for building houses or other structures only if it is necessary to attain important social interests and this need cannot be met satisfactorily on any other land." In reality, chronic food surpluses and a shift of environmental priorities to other issues have led many municipalities to pay little heed to this guideline (Hasund 1991, NEPB 1990c). Whatever the optimum land-use policy may be, such vacillation in enforcement makes for cost-ineffective policy.

Protection of the land's productive capacity from acidification, toxicity, and soil compaction was discussed in section 7.1. As for soil erosion, it is not a serious problem on a national scale and has not been the focus of much policy attention since the early part of this century, when windbreaks were established to control wind erosion in vulnerable parts of southern Sweden.

Renewable Bioenergy

Swedish energy strategy reached a critical juncture in the 1980s. Many environmental problems—local, national, and global—derive

[20] In practice, afforestation or abandonment of submarginal farmland cannot be prevented by law alone (Drake and Petrini 1985).

from heavy dependence on fossil fuels: oil spills, ground-level ozone, acidification, nonrenewable resource depletion, and greenhouse gas accumulation. In economic policy debates, the adverse effect of oil imports on the trade balance and Sweden's vulnerability to future oil shocks have been major themes. Taxes on carbon and sulfur emissions, legislated in 1990 and implemented on 1 January 1991, will significantly increase the cost of relying on fossil fuels (see Chapter 9).

After a popular referendum on nuclear power in 1980, the Riksdag voted to decommission Sweden's twelve nuclear reactors by the year 2010. Since then, it has passed additional resolutions to halt increases in carbon dioxide emissions and prevent hydroelectric dams from being built on the four remaining free-flowing rivers in northern Sweden. It is unlikely that any energy strategy could attain all these objectives without severe short- and medium-term economic consequences (Bergman 1989).

Energy conservation, induced by both high prices and clean air policies, has reduced energy consumption per krona of GDP by more than one-third since the first oil shock in 1973, and further efficiency increases are in the pipeline. Nonetheless, there is a strong national interest in developing renewable, environmentally benign, domestic energy sources. Commercial wind energy systems, for example, are being constructed in regions with appropriate airflow, and an offshore system in the Baltic is at the prototype stage.[21] Burning more residues from Sweden's extensive logging operations is a relatively low-cost option for space heat and perhaps electricity, although this would limit recycling of forest organic matter and mineral nutrients.

In this broad context, Sweden's million-ton grain surplus and the prospect of 300,000 to 1,000,000 hectares of redundant farmland have made bionergy crops a topic of great interest. Superficially, agricultural bioenergy seems to be a way to solve environmental, energy, and farm economy problems simultaneously. Several agronomic and technical alternatives have been tested and evaluated. Most R&D resources have been focused on species for burning, such as fast-maturing trees, grasses, and grain straw; grain and sugar beet feedstocks for alcohol; and fermented animal wastes for methane. In the 1980s, public funds were allocated to basic research and to prototype projects, but commercial bioenergy development faced several impediments: depressed

[21] The National Energy Board (1989) estimates that the marginal cost of additional electrical generating capacity is lower for wind-based systems than for new thermal plants, even at current low oil prices. Wind power's long-run potential is reckoned to be 10 to 20 percent of national electricity consumption.

fossil fuel prices, surplus nuclear generating capacity, immature technologies for biomass cultivation and conversion, underdeveloped systems for materials handling, and of course continuing price supports for food and feed products (Kumm 1989, SIAE 1984).

State-supported conversion of arable land to bioenergy crops is a feature of the 1990 food policy reform, analyzed in Chapter 8. We anticipate that analysis here by setting out the main qualitative effects and major uncertainties associated with conversion. Three positive effects have been claimed: first, that fossil fuel imports would decline; second, that agricultural biofuels production and processing could give an economic boost to some sparsely populated and economically depressed regions; and third, that the growing crops would absorb carbon dioxide. Although most of the CO_2 is returned to the atmosphere when biomass is converted to energy by burning, every renewable biomass option generates less net carbon and sulfur compared to equivalent fossil fuel energy.[22] In contrast to these unambiguous positive effects, the effect on nitrogen leaching varies by species and cultural methods. If grain-oilseed rotations in sandy coastal regions were replaced by perennial energy grasses or short-rotation willow (salix), for example, leaching would be significantly reduced. Finally, the impact on landscape and wildlife habitat depends on the bioenergy species, what it replaces, where it is planted, and on what scale. In an extreme case, one can imagine large, contiguous tracts of dense, five-meter-tall willow obliterating the landscape of the south Sweden plains. At the other extreme, introducing an annual energy crop such as beets into rotations might enhance visual and biological diversity.

Recycling Urban and Industrial Wastes

In the 1980s, Swedish politicans took it for granted that farmers would help recycle municipal sewage sludge. The main alternative for municipalities has been to deposit sludge in landfills, but a sharply rising incremental cost of landfills has made burning an attractive and growing option. To avoid the economic costs and environmental damages associated with these methods, most local governments agree to transport and spread sludge on farmland free of charge. As mentioned previously, the sludge and spreading procedures must meet environmental protection requirements that are quite strict by international standards (though from a cost-effectiveness perspective, the quantita-

[22] A fraction of vegetative carbon might also be added to soil organic matter.

tive limit of 4 micrograms of cadmium and 5 micrograms of mercury per kilogram of dry matter are probably inferior to an appropriate tax) (Hahn 1991, NEPB 1990c).

By the early 1980s, about 60 percent of all municipal sewage sludge was being spread on 110,000 hectares of farmland yearly, at an average rate of 1 ton/ha. Over the five-year cycle allowed by statute, almost one-fifth of arable land received sludge applications. From a sustainability perspective, nutrients and organic matter should be recycled to the extent possible, but this principle encounters two potential problems. The first one, soil contamination, appears to be well controlled through regulations. (Sludge, however, contains hundreds of chemicals, and knowledge about all the hazards they pose is far from complete.) The second, costs for transporting and spreading sludge, compares favorably with other disposal options in most circumstances (Hahn 1991, Lidén and Andersson 1989, SS 1990b).

Despite quite rigorous controls, the heavy metals problem has become a political hot potato, as discussed earlier. Sludge accounts for only 4 percent of total cadmium deposition on farmland, but on those fields where it is applied, the level (2.6 g/ha) is about twice that caused by phosphate fertilizer. The cadmium level in wheat has doubled in the past sixty years, and although the present level (50 parts per billion dry matter) is not considered a health hazard, it has received some media attention. The LRF judged the risk of an adverse consumer reaction to be greater than the benefit of free soil amendments, so it imposed a boycott in 1989, and farmland applications dropped from 60 to 10 percent of municipal sludge. In late 1991, LRF revised its position, accepting sludge that meets the 1995 target of 2 mcg/kg set by the National Environmental Protection Board.

Although the boycott was largely a public relations maneuver, it may have had beneficial dynamic effects on the economy and environment. The growing cost advantage of disposing of sludge on farmland has combined with popular support for waste recycling to strengthen government funding of projects to develop and install technology for separating industrial wastewater from household sewage and for other methods of extracting harmful substances. By one estimate, the potential joint gain to municipalities and farmers from spreading "clean" sludge on fields instead of landfilling it is about $40 million per year (Hahn 1991).

If the 1990 Food Policy Act has the expected effect of shifting some resources from food and fodder crops to energy and industrial crops, and if sludge application is concentrated on the latter, then contaminants will pose even smaller health risks (see Chapters 8 and 9).

Alternative Agriculture

"Alternative agriculture" describes a range of low- or nonchemical farming systems. The definitions used by Sweden's principal organic and biodynamic farming organizations, as well as the official certification program, affirm the value of long-term sustainability, stress the use of local and on-farm resources, and prohibit the use of synthetic pesticides and soluble mineral fertilizers. The broad-based Swedish Nature Conservation Society has declared its support for organic farming.

The government has offered subsidies for alternative production systems for four principal reasons. First, elimination of biocides is considered to be environmentally beneficial. Second, it is expected that conventional farmers will learn practical lessons from organic growers' experience with nonchemical methods of weed and pest control. Third, domestic demand for food produced by chemical-free methods is growing. Optimists, including the LRF, believe that Sweden can use its clean image to become an exporter of organic and reduced-chemical food to the European continent. Fourth, lower yields with organic methods cut food surpluses and export costs.

For a few years beginning in 1989, the bulk of public expenditure took the form of direct payments to organic farmers, an approach not common in other countries. The acreage set-aside program was modified to offer three years of set-aside payments to farmers who would contract not to use biocides or mineral fertilizers on all or part of their land. Farmers had to commit to the program in 1989 but could phase in chemical-free methods over five years, after which land had to be managed without chemicals for an additional six years.

The budget of this temporary program was US$15 million, with $5.9 million spent in the first year. Annual payments vary from $125 to $480 per hectare, according to a farm's previous crop yields and livestock intensity. (The average 1989 payment was $203/ha.) Eighteen hundred growers enrolled 42,000 hectares in the program, more than quadrupling the acreage of organically managed land and revealing the power of economic incentives at a time of declining farm income. The extent of certified organic farming—1.45 percent of cropland—is almost certainly the greatest in any industrial nation (Haxsen 1990).[23]

Financial incentives are complemented by advice from nine organic extension specialists. Complementary education and training activities have been developed for county extension agents. The University of Agricultural Sciences has established a Department of Ecological Agriculture and financed several research projects in reduced-input farming; in addition, a five-year, $2.7 million organic farming research investigation was launched by the Swedish Council for Forestry and Agricultural Research. Government activity has concentrated on the production side, with market development largely left to farmers' associations.

It is understood that the changeover from conventional to organic farming has complex economic and environmental effects. In most cases it entails less use of off-farm inputs and more on-farm nutrient cycling, new rotations and commercial enterprise mixes, and substitution of mechanical and cultural methods for chemical biocides. The effects depend critically on the specific production systems and the quality of farm management. In practice, the majority of converts have been grain producers attempting to continue their specialization without chemicals.

Elimination of pesticides from the local environment, the workplace, and farm products is a clear social benefit. Swedish evidence also indicates that fossil fuel intensity per hectare and per ton of output is lower in organic farming than in conventional systems, with increased energy use in field operations (especially weed control) outweighed by energy savings from reduce nitrogen application (Brorsson 1991). The implications for Sweden's most serious agroenvironmental problem—nitrogen leaching—are less clear. Dairy operations are central to many organic farms, and organic milk production is a nitrogen-intensive system. Organic soil management centering on legumes and animal manures does not eliminate nitrogen leaching; indeed, under certain conditions, organic methods leach as much as conventional systems (Granstedt 1990).[24]

The overall environmental effect of the government's support measures is positive, but their cost-effectiveness is dubious. If the principal objective is to achieve the national environmental goals of reduced leaching and pesticide use, one must be skeptical about the use of

[23] Much of the information on this program comes from a 1990 interview with Gunilla Wennberg, National Board of Agriculture.

[24] Much of this information was provided by Maria Wivstad, Swedish University of Agricultural Sciences, SLU Info.

subsidies averaging more than $200 per hectare to affect less than 1.5 percent of cropland. Furthermore, a six-year contract does not guarantee chemical-free management over the long run. In our view, direct payments would be a sound use of public funds only if the participating farmers created significant learning benefits for conventional farmers not enrolled in the program. In reality, the program did not systematically use the enrolled farms for field experiments or demonstrations. If export outlets for organically produced foods were developed, such experimentation and demonstration roles could have a big economic payoff (Brorsson 1989).

7.4 Protecting Animal Rights

The spread of intensive livestock production has brought several interconnected aspects of animal husbandry, especially health care, feeding regimes, confinement conditions, and slaughter procedures, under criticism from animal-rights organizations and under scrutiny from the mass media. Concern about the safety and quality of human food also colors the politics of animal husbandry. As mentioned, the revered children's storyteller Astrid Lindgren has led the crusade for "healthy and happy pigs."

Limited knowledge, not to mention vested interests in the status quo, hinders effective policy design. There are no simple formulas to translate knowledge from animal sciences into quantitative measures of how livestock management practices affect animals' physical wellness. Getting an analytic grip on the sociopsychological well-being of domesticated livestock and its link to husbandry practices is fraught with even greater ambiguity. Two special problems arise in deriving policy prescriptions from scientific analysis. First, there is strong evidence that four-legged animals' well-being is greatly affected by the quality of husbandry—the subtle relationship between animals and their human keepers. Truly considerate care cannot be guaranteed by mechanistic regulations on feeding and housing or by investments in training stock keepers. Second, though animal-rights organizations, and the new regulations, demand that livestock be reared in "more natural conditions," what exactly "natural" means is problematic for species that have been domesticated for millennia, selectively bred for centuries, and genetically manipulated in recent times (Hellstrand 1988, Rowinski and Johnsson 1990).

Notwithstanding these rather awesome knowledge gaps, Sweden's animal-protection law was thoroughly rewritten in 1988, with special emphasis on commercial livestock (experimental and recreational animals were largely excluded). The guiding principle is that domestic animals should be ensured a healthy, low-stress environment with opportunities to behave according to their natural ways. Cows are thus given the right to graze at least three months a year, hogs must be allowed to follow socializing instincts, and laying hens should be able to perch and sandbathe.

The rules are immediately applicable to new stables, but for existing structures target dates are set at various points in the 1990s. A good housing environment is one major goal. General rules apply to stall gases and dust, mechanical noise levels, and minimum confinement space.[25] Barns must admit daylight and have appropriate artificial lighting. Animals are entitled to superintendence daily and fresh water twice daily (continuous water supply is recommended). Standards have been set for the nutrient and fiber content of feeds, and the earlier ban on growth regulators and routine use of antibiotics continues in effect (NBA 1989b).

Numerous special rules and exceptions for different species make it difficult to generalize about the new law or evaluate it as a whole. For instance, land-scarce research farms such as Alfa-Laval's operation outside Stockholm have been exempted from the three-month grazing rule. There is no obligation to include coarse fodder in cattle rations, despite its known benefit to their digestive systems. A sow can no longer be bound in place longer than one week during farrowing, and piglets must be kept with their mother for the first four weeks, yet she need not be allowed to follow her nest-building instinct. Laying hens must be kept in free-range confinements, but there has been no ruling whether a prototype cage designed for ten hens and constructed with perches, nests, and sandbaths will qualify (NBA 1989b, Pettersson 1990).

Evaluation of this complex and detailed regulatory package is beset by uncertainties about how particular rules affect animal well-being and by unresolved questions about how this goal should be balanced against economic costs. A prime example is the requirement that all laying hens be kept in floor-range housing by 1999. A comparison has been made of commercial and experimental housing systems used in several countries. In general, the unit cost of eggs produced under

[25] Air quality standards are: ammonia, maximum 10 ppm; carbon dioxide, 3,000 ppm; hydrogen sulfide, 0.5 ppm; and organic dust, 10 mg/m^3 (NBA 1989b).

floor-range conditions is 20 to 30 percent higher than in the three- or four-hen cage systems prevalent in Sweden and most other industrial countries. The main cost inflator is less efficient feed conversion, but depressed egg production per hen and greater labor intensity also raise unit costs. With the limited technical options currently available, these costs will be borne in some combination by consumers, taxpayers, and producers when "natural" housing comes on line. The political choice is further complicated by serious doubts about whether natural housing really improves hens' well-being. Greater freedom to roam, perch, and sandbathe comes at a cost of increased parasitic diseases and more deaths and injuries from fighting. The prototype ten-hen cage mentioned above is admittedly not natural, yet it might help limit both the economic and ethological conflicts (Rowinski and Johnsson 1988).

Similar contradictions arise with the requirements for more natural hog confinement. Here unit costs are likely to rise 10 to 30 percent. The main dysfunction is an increase in deliberate and accidental killing of piglets by unconfined sows (Rowinski and Johnsson 1990). A recent study compared Sweden's highly detailed requirements with Denmark's more general regulations and concluded that the Swedish approach will raise relative wholesale cost by 5 to 8 percent. The principal contributing factors are higher housing costs, slower weight gain (growth hormones are banned), increased labor requirements, and a more expensive slaughter process (Malmström 1991).

Increased stall size and other requirements for housing cows will probably have capital costs of US$25 million nationally. Combined with manure storage requirements, this expense will likely accelerate the exodus of small dairy farms in the 1990s, with negative impacts on landscape amenities, biological diversity, and rural economic vitality in sparsely populated regions (Hasund 1991, NBA 1990c).

In 1988, government arm-twisting got livestock producers to accept price penalties on carcasses when inspectors find evidence of illnesses and physical malformations caused by poor husbandry. The penalties have been set higher than the medical costs and yield losses normally caused by the health defects to create a strong incentive to curb managerial malpractice. Catherine Belotti (1991) estimates the annualized cost of bovine illness to be at least $50 per cow in veterinary fees, medications, and lost yields. She concludes that the penalties will indeed make it profitable to improve husbandry.[26] On top of the penal-

[26] Penalties range from US$1 to $20 per physical defect for slaughter cows. Our information on slaughter penalties comes from Nils-Gunnar Nilsson of the National Food Administration.

ties, producers who deliver many defective animals are subjected to criticism and advice from the extension service.

Unavoidably, strict animal protection laws have implications for rules governing importation of livestock products. If a certification regime is implemented, so that imports of products from animals raised under less humane conditions are either prohibited or subjected to countervailing import duties, then Swedish livestock farmers and processors will be able to pass part of the cost on to consumers. The great majority of Swedes view humane treatment of livestock as a moral obligation, but in practical terms, they can be expected to cut back consumption if retail prices rise.[27] Without compensatory border protection, Swedish producers would suffer a major loss of market share for products other than fluid milk; many would be forced out of business. Perversely, the intended improvement in animal well-being would also be nullified if products from animals raised in less humane conditions entered freely.

In sum, preliminary analysis indicates that there will be a sizable price to pay when animal-rights and slaughter-penalty statutes are fully enforced. But critics within the animal-rights movement are not convinced that the 1989 law will even bring humane and natural conditions for farm animals (Jensen 1990). On the positive side, the lead times built in to most new requirements allow time for, and focus political pressure on, a range of constructive actions: more comprehensive scientific analysis of conditions affecting animal well-being, development of new husbandry techniques, farmer training, and modification of regulatory details when evidence of deficiencies emerges.

One encouraging example of the technological innovation induced by new regulations is "Power Fodder," developed by an LRF branch organization and introduced to the Swedish market with a public relations campaign stressing its benefits to livestock and the environment. The incentive to develop a less protein-rich diet was strengthened by the regulations penalizing animal illnesses, curbing ammonia in barns, and banning preventive antibiotic dosing. Power Fodder uses essential amino acids produced by microbial action to create a protein that is more efficiently absorbed than standard feed rations. The lower total protein content reduces the frequency of respiratory and intestinal diseases while improving general animal well-being. The new fodder also improves feed-conversion efficiency and slightly increases fertility

[27] Demand is elastic for beef and just below unity for pork and poultry (Holm and Drake 1989, Silvander and Drake 1990).

in cows and sows. Air and water quality will be improved by the reduction of nitrogen compounds in animal wastes. And fertilizer applications on feed grain and ley crops can be cut back because high protein content will bring less of a price premium. Finally, the new fodders should reduce Sweden's import bill for protein feeds; they may even become a green export in their own right (LRF 1990a).

7.5 Five Years of Action: Summary Observations

Swedish politicians acted boldly in the latter 1980s to limit modern industrial agriculture's negative effects on the environment, human health, and animal welfare and to affirm agriculture's noncommodity benefits to present and future society. These are tangible signs that agriculture is taking on a new meaning. Nevertheless, the array of initiatives described in previous sections was not inspired by a clear, holistic vision of the agricultural future. Nor do they sum up to a coherent green strategy, paying close attention to complementarities and conflicts among the many policy measures.

Some initiatives, such as animal-protection regulations, have been ad hoc reactions to mounting political pressure. Several other measures would not have been undertaken, certainly not with the same timing and financial commitment, but for the overriding goal of cutting production surpluses. And an array of instruments introduced piecemeal over several years inevitably contains inconsistencies. Thus, subsidized spruce planting was used as a supply-control measure, even as the NOLA supports promoted the opposite goal of protecting open and varied landscape. In view of how the policy evolved, there are actually comparatively few such inconsistencies. The most obvious side effects were usually identified in the policy design process, with public scrutiny and debate often playing a key role.

The top-priority environmental problem, nitrogen leaching, prompted a many-sided policy response that has had substantial positive impact, but the initiatives are more of a grab bag than a carefully crafted package. Cost containment and potential synergies among different instruments seem to have been given little thought.

Nearly all the initiatives sketched in this chapter have made some contribution to their stated objectives, a possible exception being free-range housing for laying hens. Few measures have had unintended negative effects on other green objectives (Table 7.5). The adverse landscape and nature conservation effects of conifer plantations and

Table 7.5. Classification and evaluation of green policy instruments

	Environmental pollution		Nature conservation		Sustainable resource use		Animal wellness	
Policy instrument	ga	ce	ga	ce	ga	ce	ga	ce
Fertilizer tax/levy	+	+	(+)		+	−		
Manure storage capacity	+	−	(−)[a]		+	−		
Manure spreading regulation	+	+						
Catch/cover crop requirement	+	+	(+)		+	+		
Livestock density limit	+	?	(+)		(+)		(+)	
Biocide tax	+	±	+	±	±	?		
Biocide certification	+	±	+	±	±	+		
Set-aside modifications								
Catch crops, ley	+	+	(+)		+	+		
Afforestation	+	−	−[b]		±	?		
Organic farm support	+	−	(+)		+	−	(+)	
NOLA support			+	+	(+)		(+)	
NYLA support	(+)		+	−	±	−		
Animal-protection law	(+)[c]		−[a]				+[d]	?
Price penalty/sick hogs	(+)[c]						+	+

Source: Authors' assessment, based on published references, empirical evidence, and own deductions.

Note: ga = goal attainment, ce = cost-effectiveness, + = positive direct effect, − = negative direct effect, ± = mix of positive and negative effects, ? = direct effect unclear, ( ) = indirect or unintended effect, *blank* = no or negligible effect.

[a]Capital costs of compliance are likely to drive out some small dairy producers, especially in disadvantaged regions.

[b]Afforestation can have a positive effect on landscape, depending on region and species plants.

[c]Likely reductions in nitrogen from animal wastes.

[d]Some components of the law have doubtful benefits for animal well-being.

of the manure storage and animal-protection directives are minor exceptions.[28] The implementation timetables provide some flexibility to reconsider such measures before they cause problems on a large scale.

Positive goal attainment does not mean that instruments maximize goal attainment or minimize cost. A uniform fertilizer tax, for example, is target-inefficient and has unnecessarily high opportunity costs. Table 7.5 summarizes our provisional conclusions about effectiveness across the spectrum of green instruments, premised on the agricultural policy of the latter 1980s and based on a mix of deductive logic and limited empirical evidence. It is important to keep in mind that the social benefits of several measures, such as fertilizer taxes, NYLA payments for landscape restoration, and the organic farming subsidy, are increased by their contribution to grain-supply control. Both the economic and environmental rationales for these measures would be weaker in a reformed policy regime with lower grain prices.

The experience gained between 1985 and 1990 warrants several further generalizations. First, many, if not most, measures were adopted before their environmental and economic consequences had been well researched, much less tried out in pilot studies under realistic commercial farming conditions. The targeted 75 percent reduction in biocide dosage by 1995 and some of the livestock housing standards have little scientific justification.

Second, there remains a strong bias toward regulatory requirements, even when market-like solutions (taxes, marketable permits, subsidies) would be less costly. The slaughter penalty on sick animals, the new risk-based tax formula for biocides, and NOLA landscape rents are notable exceptions. When economic instruments are used, they are not always well conceived, as in the case of the uniform nitrogen tax. The economic inefficiency of command-and-control approaches will be even greater once grain surpluses are eliminated.

Third, most measures to reduce negative externalities attempt to make polluters pay. In practice much of the cost will be passed forward to food consumers, so long as high import barriers remain. The most blatant contradictions of the Polluter Pays Principle are tax-financed grants for manure storage facilities and cost-sharing for planting catch crops. Among farmers, the equity principle that individual polluters should pay in proportion to their share of damages is not consistently applied.

[28] The capital costs to meet manure storage and animal housing requirements are sufficiently burdensome that some small farms will drop out of dairy production. As noted, dairy production is closely tied to landscape quality in economically disadvantaged regions.

Fourth, green measures compensate some producers for providing public goods, yet the majority receive no such compensation and the payment levels typically fall well below our estimates of social benefit.[29] Compensation for maintaining *hagmark* is too low, but for converting to organic farming it is too high.

Fifth, setting compliance deadlines five to ten years in the future allows politicians to take popular green stands today while deferring most of the budgetary cost and socio-economic "pain" to the future. Of course, deferred target compliance dates also have the economic benefits discussed earlier: they make it easier for producers to learn new production methods, for researchers to develop efficient technical and managerial solutions, and for government to fine-tune regulations as new information uncovers defects in the original measures. An example of this administrative learning curve is the ongoing review of livestock housing requirements.

Finally, some of the green measures are experienced by many Swedish farmers as psychological and economic burdens, imposed by urban politicians who possess little understanding of or sympathy for agriculture.

Taken together, the green initiatives have contributed significantly to green policy goals, but with weaker goal attainment and at a higher cost than could have been achieved with a better-crafted set of measures. Realistically, a pluralistic democracy cannot be expected to attain the neoclassical economist's ideal of "Pareto efficient" policy design. If future policymakers uphold Sweden's reputation for pragmatic fine-tuning of policies, Swedish citizens can expect increased greening per krona over time.

[29] Admittedly, the policy objective in some cases was to compensate farmers for costs or losses, not to pay for provision of public goods.

8

Food Policy Reform: A Pale Green

At the same time as environmental problems and green demands were being translated into piecemeal policy measures in the latter 1980s, the old policy's economic deficiencies and a broad neoliberal shift in Swedish politics brought mounting pressure against the entire agricultural policy apparatus. But how does a nation go about dismantling an enormously complex, half-century-old regulatory system? How can longstanding farmer entitlements be phased out in a socially acceptable way? Is it possible for a market-oriented reform to satisfy green demands? Is such a reform really the magic bullet its backers promise, improving economic efficiency, consumer welfare, and environmental performance all at once? This chapter responds to these questions.

8.1 Negotiating a New Policy

In the latter 1980s, convergent economic, political, and ideological forces prompted a multiparty effort to break out of agricultural policy inertia. These were years of fundamental political rethinking and party realignments in Sweden. The state's intrusive regulatory apparatus came under heavy criticism, and the Social Democratic government launched preemptive market-oriented reforms in several areas. Within this big picture, the fossilized agricultural policy's deteriorating goal attainment and escalating cost, coupled with its environmental contradictions, drove the reform process. A national consensus on the need to curb runaway government expenditure and fight inflation helped focus attention on agricultural policy, since it worsened both problems. We contend that the agricultural components of Swedish food policy actually received more blame for exceptionally high food prices than the facts justified. Farm receipts comprise only about one-fourth of retail food expenditure, and most of the 1980s food inflation can be traced to higher costs and profit margins in the food distribution system, where weak competitive forces and wage inflation had a major impact on retail prices. Policy was implicated too, via import protection for first-stage food processing and urban zoning ordinances that created entry barriers to new retail food competition. Ironically, the LRF's support for the food processors may have exposed farmers to a reform that could have been delayed or weakened.

In a comparative perspective, we observe that several inhibiting forces that perpetuated farm policy inertia elsewhere operated less powerfully in Sweden. Swedes do not have the vivid memories of wartime starvation that motivate Finnish farm policy, the oil revenues that finance Norway's heavy farm subsidies, or the combersome institutional arrangements that long obstructed reform of the EC's Common Agricultural Policy.

The Policy Process

In 1988, a Swedish state commission was charged with designing a fundamentally new agricultural policy for the first time since the 1930s. The initial proposal was prepared by the Food Policy Working Group (Livsmedelspolitiska Arbetsgruppen, or LAG). The task assigned it by the Social Democratic government was straightforward. It should propose ways to deregulate the domestic food market; devise directed measures for food security, regional economic balance, and environmental protection; and prepare the way for reduced border protection, at a pace to be determined by the success of GATT negotiations (MOA 1989). The working group was composed of Riksdag members from the six parties, with no formal participation by representatives of interest groups or state agricultural agencies. Its secretariat consisted of career civil servants from outside the agricultural establishment but with close ties to the Ministry of Finance and its liberal economic thinking.[1] These tactics were clearly intended to curb the influence of the iron triangle.

[1] The LAG's secretary, Michael Sohlman, held the position of state secretary in the Ministry of Agriculture, but his previous appointments had been in other branches of government.

The secretariat produced a thorough analysis of the old policy's shortcomings from a neoclassical economic viewpoint and a social democratic political viewpoint. Although negative environmental externalities such as nitrogen leaching were rhetorically highlighted, the stress was on economic policy failure. The proposal was cast in environmentally friendly language, but its anticipated green benefits were largely incidental to economic deregulation. In particular, lower farm-gate prices would weaken incentives for grain monocultures and intensive use of fertilizer and biocides.

The LAG proposal had four core elements: a market-orientation principle, a new goal structure, permanent policy instruments, and transitional measures. Although the LAG's recommendations were extensively amended during the legislative process, these four elements were retained in each successive version of the policy.

The market-orientation principle held that food production and distribution should be exposed to the same competitive conditions as other sectors, with producers compensated only for production corresponding to demand. State intervention is justified only if free market supply fails to meet demand *and* production has some public goods characteristics. Directly targeted measures, not market regulations and price manipulation, should be used to secure public goods. Food security, environmental protection, and regional economic vitality are the core public goods.

The new goal structure redefined traditional policy objectives and juggled their relative priorities. In brief, consumer well-being and food security were accorded highest priority. After more than forty years as a basic principle, fair compensation to farmers was no longer construed as an end in itself, but as means to other collective ends.

The proposed permanent policy regime recommended that domestic commodity markets be deregulated, with export subsidies eliminated and market-determined prices no higher than the market clearing level. A degree of import protection from other nations' subsidized exports should be maintained until multilateral trade liberalization could be achieved, and variable import levies should eventually be converted to fixed-rate tariffs.

For national security and other lower-priority reasons, Norrland farmers should receive direct income support to offset cost disadvantages caused by market deregulation; and two-thirds of oilseed acreage should be maintained temporarily through price-supports and contract production.[2] On the basis of a cost comparison of alternative methods, it was recommended that food security be maintained with less peacetime food self-sufficiency and farmland, but larger stockpiles of critical farm inputs and staple foods.

Nature conservation would be promoted by spending more on long-term land-management agreements, analogous to NOLA contracts. Additional investment funds were proposed for rural development in regions where the new policy would intensify economic recession and depopulation.

Deregulation implied removal of the fertilizer and biocide levies that had been used to finance exports of excess grain, but the working group deferred a decision on green input taxes to the parliamentary commission on economic instruments to achieve environmental goals, mentioned above. LAG did clearly state, however, that incentives to economize nitrogen should not be weakened.

The LAG report's transitional measures included three years of compensation to grain producers for lost income. They would be compensated at a declining rate, based on their past acreage and yields of regulated crops. LAG did not recommend any compensation for permanent conversion of cultivated land to forest or nonregulated crops. It did, however, propose that the state purchase financially insolvent farms at their pre-reform market value and resell them on the open market (MOA 1989).

The LAG proposal was a product of the working group's Social Democratic secretariat. The other political parties added appendixes expressing their reactions. All the parties agreed that long-term domestic market deregulation was right and that unilateral elimination of import protection was wrong, but there were extensive disagreements about the transition process and methods of compensating farmers. The nonsocialist parties favored a longer period of transitional compensation but were split on the issue of how to compensate farm owners. The Center and Moderate parties (representing most farmers) argued for a slower, more gradual cut in support prices instead of compensation based on acreage. The Liberals, in contrast, urged a permanent, flat-rate acreage subsidy (SP 1990).[3]

[2] It was estimated that import restrictions alone would sustain only about one-third of oilseed production.

[3] The rationale for a permanent acreage payment was to compensate farmers for their contributions to food security and landscape values. This method is similar in some ways to the MacSharry Plan for reform of the EC's Common Agricultural Policy, discussed in Chapter 9.

There were also differences over the proposal's environmental component. The Center and Green parties proposed greater expenditure to protect the most valued landscapes, but they wanted to spend only about one-tenth as much as the Liberals for permanent land compensation.

In the area of input taxes, the Greens at one extreme advocated a 300 percent tax on nitrogen fertilizer, with part of the revenue used to underwrite conversion to organic farming. The Left party also urged more grants for conversion to chemical-free farming. The Moderates, at the other extreme, opposed any input tax increases, arguing that they would not be cost-effective once lower farm prices eliminated excess production. No other party supported cuts in nitrogen or biocide prices.

In view of the radical changes contained in the LAG proposal and the political parties' differing responses to it, it might seem improbable that a majority could be forged around a reform package. But, with the possible exception of the Center party, all were determined to be rid of the ineffective old policy and all were prepared to pay a sizable short-term price to buy it out. The Social Democrats had an interest in pushing through a proconsumer reform without bearing sole responsibility for any negative consequences. It was also in the bourgeois parties' interests to reach a consensus that would avoid later intracoalition conflicts, in case they should win the 1991 general election and form a government.[4]

In the debate's early stages, LRF refused to discuss any alternative to the existing high-price policy. But when the multiparty support for deregulation became clear, LRF shifted its focus to lobby for a permanent land bank, with acreage payments to compensate farmers and sustain asset values. Payments would be based on past yields, as in the 1980s set-aside program. LRF proposed supplementary acreage payments to maintain production and open landscape in less fertile forest regions. It continued to oppose any price deregulation, justifying its position in the case of milk with the argument that competitive markets would bankrupt thousands of small milk producers. In remote areas, especially, the collapse of farming would undermine food security, regional balance, rural incomes, biological diversity, and the cultural-historic landscape.

Finally, LRF pushed hard for a major public commitment to energy

[4]The SAP, its voter support declining, was also anticipating the 1991 election. A left-center-right consensus on food policy would reduce the chance that a proconsumer reform could be undone if the SAP should lose the election.

crops and biomass energy plants. LRF's economic motive was to maintain intensive production systems and high land values, although its public pronouncements stressed Swedish energy independence, rural economic vitality, and environmental benefits (less nitrogen leaching and fewer carbon emissions or nuclear power risks). The agroindustrial lobby participated relatively little in the public debate, often allowing the LRF to represent their mutual interests (LRF 1990b).

Environmentalists warned of the LAG proposal's negative effects on sustainable agriculture, biological diversity, and the cultural landscape. In its remiss response, the Swedish Society for Nature Conservation (SNF) supported Liberal party and LRF proposals for permanent and universal acreage payments, though it urged that rental payments be varied according to the land's public goods contribution. SNF also urged higher fertilizer and pesticide taxes (SNF 1990).

Compromise and Consensus: Legislation Takes Shape

The government's legislative bill, submitted early in 1990, took many of the criticisms of the LAG report into account. It declared the food policy's objectives in these terms:

> A principal goal is that food policy shall be in accordance with the general objective of economizing society's total resources. Food shall comply with hygienic and informational standards set by food legislation. Simultaneously, the policy shall contribute to a well-balanced diet and thereby to improved public health. Food price developments shall be reasonable in relation to price developments for other goods and services. Consumer's choices shall govern production. Consumers shall have good opportunities to choose among foods of different types, regarding taste, origin, production methods, or degree of processing. Food security shall be based on the sectors' peacetime resources and their capacity for adjustment, complemented by specific preparedness measures. The goal is to make the nation's food supply secure during crises and in war. In the choice of land-use and production methods, account must be taken of the demands of a good environment and farsighted economizing of natural resources. The environmental goal is to safeguard a rich and varied agricultural landscape and to minimize the negative environmental impacts of agriculture. Fair compensation to producers for products and services is a prerequisite for other goals to be realized. The food policy shall contribute to regional equalization of employment and welfare. (GP 1990b:49; authors' translation)

The bill's economistic and proconsumer thrust is clear, yet in several ways it took a stronger proagrarian stance than the earlier LAG proposal: financial support was included for land-use conversion to wetlands, forest, and energy crops; more funds were earmarked for landscape maintenance; income compensation to farmers was increased by one-third; and some transitional programs were stretched out to five years. The projected budgetary cost of the government's proposal was US$2.25 billion over five years, or more than $23,000 per farm (MOA 1990).[5]

The Riksdag's Permanent Agriculture Committee, seeking unanimity, engineered further compromises in the spring of 1990. The Social Democrats accepted even more generous short-term farmland conversion payments, balanced by smaller direct income transfers; and the Center and Moderate parties accepted a greater degree of long-term market deregulation than their farmer constituents wanted. Overall, the LRF was fairly successful in cutting farmers' economic losses and ensuring the perpetuation of many forms of state intervention well into the future. Only the Green party refused to vote for the final legislation submitted by the Riksdag's Permanent Agriculture Committee. The Greens insisted on stiffer fertilizer and pesticide taxes (with the revenues rebated to farmers as an acreage payment) and more subsidies to organic farming. Without doubt, the new food policy was more a reflection of consensus on the old policy's failings than a shared answer to the agrarian question. Like its 1930s antecedent, it was a cow trade.

8.2 1990 Food Policy

Although the food policy legislation underwent further modifications in its passage from the government's bill to its final form, its central provisions were clear. The two basic components are new permanent instruments consistent with the market-orientation principle and measures to accomplish the transition to market-orientation "in a socially acceptable form." Commodity prices will no longer be set by government after negotiations. They will be gradually reduced to the domestic market clearing level (or even lower, depending on the outcome of GATT negotiations), and a vast number of detailed regulations will be terminated. Permanent interventions are intended to be transparently linked to their objectives.

[5] Not all this total would reach the farmers as compensation.

During the transition phase, the policy remains quite complex and in some ways ambiguous, so only its core elements are highlighted here. Since several of the transitional measures have an impact on longer-term economic and environmental outcomes, they are described in some detail.

Domestic Market Deregulation

In the long run, domestic prices, with a few exceptions, will be determined by forces of supply and demand. Collectively financed exports are being phased out by 1993/94, though import levies continue to insulate producers from international competition, pending a GATT treaty. Support prices are being cut stepwise over four years, with the nominal floor price of grains, for example, dropping by 35 percent, from US$0.23/kg in 1989/90 to $0.15 in 1993/94. The 1993 price is lower than the average international price of $0.16/kg in the latter 1980s[6] (NAMB 1990). The principal exception is the Norrland, where farm prices continue to be adjusted upward to cover production costs.

Exceptions to price deregulation during the transition period include a state-supported contract system for oilseeds, a simplified scheme to equalize prices for milk channeled to different end uses, and a three-year declining beef-export subsidy, to accommodate an expected meat surplus as dairy herds are sold off.[7]

Permanent Targeted Supports

Directed public assistance is given for three main purposes: preservation of cultural-historic landscape and biological diversity, food security, and regional development. The legislation states emphatically that such support should be transparent, that is, directed to a specific target and financed out of the state's budget, not by price supports.

Aggregate payments for landscape preservation were roughly US$17 million in 1990/91, rising to a steady-state $42 million/year by 1992/93. This compares with about $7 million/year under the previous NOLA and NYLA programs. In recognition that this is an environ-

[6] Inflation will further erode the guaranteed price. At a 5 percent inflation rate, the real wheat price in 1993 would be only $0.13/kg in 1990 dollars.

[7] Milk price differentials will be reduced by using levies on profitable cream and fluid milk processors (primarily near large population centers) to subsidize dairies that process less profitable solid milk products. For the duration of this cross-subsidy, processors will be able to pay the same fluid milk price to all farmers (although legally they may vary prices). The meat-export subsidies last for three years and have ceilings of US$33.3 million in 1991/92, $25 million in 1992/93, and $16.7 million in 1993/94.

Table 8.1. State budget for targeted food-security measures (in millions of 1990 US$)

Measure	1988/89	1989/90	1990/91
Input and food storage	20.8	24.8	31.9
Norrland support	83.5	99.2	104.8
TOTAL	104.3	124.0	136.7

Source: GP 1990b.

mental program, not a covert production stimulus, it is administered by the National Environmental Protection Board and participants are selected by the regional environmental authorities. At the owner's initiative, multiyear management agreements may be contracted for arable land or natural pasture. There must be good reason to believe that the land is threatened with abandonment or afforestation, and parcels are ranked according to a complex system that includes biological, cultural-historic, and scenic-recreational benefits.[8] The payment level, however, is not based on the magnitude of social benefits, but rather on the magnitude of the land's (negative) economic rent in commercial terms. A separate NOLA program for the biologically richest lands remains; its budget was more than doubled to $13 million starting in 1990/91 (SP 1990).

To stem regional economic decay and maintain food security, assistance to farms and small-scale food processing industries in the north continues. Direct investment assistance to Sweden's sparsely populated regions has been increased by 43 percent ($17 million/year), and farms that combine agriculture with other productive activities are eligible to participate. Agroindustrial ventures in these regions are backed by grants for research, extension, and direct investment.

The new food-security strategy combines continued production on economically competitive farms, mostly in south and central Sweden, with two targeted measures: Norrland price supports, and expanded storage of staple foods and critical imported inputs (Table 8.1).

Transitional Measures

In addition to the temporary market interventions for oilseeds, milk, and beef, described above, the key transitional measures are the following:

[8] Scenic, recreational, and touristic values have secondary priority and are decisive only if the local budget is fully allocated and competing pieces of land are rated equal according

- Temporary income support payments
- Payment to terminate dairy operations
- Payment for "permanent" conversion of land out of price-supported crops
- Special payments for wetland construction and deciduous tree planting on arable land
- Financial assistance to heavily indebted farmers
- Transitional extension services.

Temporary Income Support. Producers of price-supported crops receive three years' compensation based on a formula combining past acreage and yields. Land previously enrolled in the set-aside program is also eligible. Payments over the three years average $180, $150, and $120 per hectare, which is equivalent to 110 percent, 92 percent, and 74 percent respectively, of the average (implicit) rent on Swedish farmland in the latter 1980s.[9] Translated into the equivalent grain price, the payment amounts to US$0.25 per kilogram in 1991/92, or more than the previous support price. The objective is to prevent a socially unacceptable depression of farm incomes as crop prices are cut. The aggregate budget cost, expected to be about $750 million, is partially offset by lower export subsidies. Past direct payments per cow of $167 each year also continue in effect for three more years. For milk producers, including a few goat farmers, these payments sum to about $100 million per year for three transitional years (Jonasson 1991b).

"Permanent" Land Conversion. Owners of arable land taken out of regulated crops were paid a lump sum averaging US$1,500 per hectare if the shift was made in 1991/92. The payment declined to $1,000 in 1992/93 and $667 in 1993/94. Compensation is in part fixed and in part proportional to past yields; it takes the form of a conditional interest-free loan, to be forgiven if the new land use is maintained through 1996.[10] In effect, any permanent form of conversion that farmers could invent was potentially eligible. Some of the eligible uses are forest, perennial energy and industrial crops, annual energy and

to biological and cultural-historic criteria. The detailed criteria are representativeness, uniqueness, species richness, presence of threatened habitat or cultural elements, historic continuity of cultivation, present cultivation conditions, integrity (versus fragmentation) of the preserved features, public accessibility, beauty, and educational and research value (NEPB 1991).

[9] In 1990/91, payments ranged from $120 to $240/ha, 40 percent in the form of a flat payment per hectare and 60 percent proportional to past yields. This parallels the formula used in the 1980s set-aside program (GP 1990b).

[10] The range of payments fór land converted in 1991/92 was $970 to $1,980 ha.

industrial crops if grown under contract to a buyer, niche crops such as flax, and extensively grazed pastures.[11] If a forest or bioenergy planting proposal conflicts with nature conservation goals, the conversion plan may be rejected and financial support withheld. (This issue has become controversial in some parts of Sweden.) The policy's unofficial target is to withdraw 500,000 hectares, and applications for more than 350,000 hectares were approved in the first year alone, with compensation totaling $500 million (SBA 1991c).

A premium is paid for establishing wetlands and planting either deciduous or bioenergy forest. The first two uses are considered especially valuable for biological and amenity values. Although the legislation is not explicit about the motivation for subsidizing bioenergy plantations, it seems to be based on reducing dependence on imported fossil fuel. The high level of payments per hectare has meant that the $67 million allocated to these purposes covered only about twenty thousand hectares.[12]

In the area of dairy operations, $90 million will be used to buy out herds, and the voluntary pension plan for older dairy farmers will continue until 1996. From 1991 through June 1993, the state also paid dairy operators to cut milk production ($0.06/kg in 1991 and $0.04 thereafter).

Financial Assistance and Special Extension Services. Several aids are available to help farmers reconstruct or liquidate their businesses if they face severe economic problems directly caused by deregulation. One mechanism is a 2.5 percent interest rate subsidy on consolidation loans (to a maximum of US$8,300/year). In extreme cases, the state may purchase farms from bankrupted or retiring operators at the prereform market value and resell at the current lower price. (There have been very few cases as of 1993.) According to the newly created Swedish Board of Agriculture, 1.5 percent of farms are expected to face foreclosure and another 3 to 5 percent will need reconstruction assistance. Tenants are guaranteed the right to renegotiate leases at lower

[11] For consistency, the prohibition against land conversion under the Land Care Act was removed. Farmers are still obligated to report any planned arable land conversion eight to twelve months in advance, so authorities will have time to offer a management contract or take other steps in the public interest. Sheer abandonment of land remains illegal, but this cannot be fully enforced. The Silviculture Law states that land suitable for silviculture should be afforested if not used for another purpose.

[12] Farmers may receive up to $3,300/ha to establish two hectares of wetlands plus $1,150 for additional land; deciduous forest is eligible for a maximum of $2,300/ha plus up to $670/ha for fencing to keep out browsing mammals.

Table 8.2. Projected budget costs of the food policy reform (in millions of 1991 US$)

Policy instrument	Fiscal year 1990	1991	1992	1993	1994
Transitional supports					
Income compensation	348	233	167	0	0
Land-conversion premium	600	450	50	0	0
(wetland/deciduous forest)	33	33	17	0	0
Meat-export subvention	33	25	17	0	0
Dairy buyouts	60	0	0	0	0
Dairy pension plan	12	8	?	?	?
Bankruptcy assistance	3	3	3	3	3
Entry farmer assistance	3	?	?	0	0
Permanent supports[a]					
Food security (storage)	51	33	33	33	33
Norrland support	110	110	110	110	110
Landscape contracts	17	33	42	42	42
NOLA (biological richness)	8	8	8	8	8
Animal grants	100	100	100	100	100
Regional development	21	17	17	17	17
Total committed expenditures	1,574	1,278	755	413	413

Sources: GP 1990b, MOA 1990a.
Note: ? = no estimate available.
[a]Several of these supports existed before 1990.

rates as declining commodity prices drive down the market value of land.

Transitional extension advice from experts in finance, production, and other fields is available, on the premise that many farmers need retraining to remain viable in the new economic conditions (SBA 1991a). Under Sweden's active labor market policy, farm owners, tenants, and wage earners who leave agriculture qualify for generous unemployment compensation plus a comprehensive set of career counseling, job search, training, and relocation services.

The official estimates of the new policy's budget costs in its first five years are listed in Table 8.2. The effect of concentrating income compensation and conversion payments in fiscal years 1990 and 1991 was to push Sweden's producer subsidy equivalent to an all-time high of 63 percent in calendar year 1991.

8.3 Economic Assessment of the New Policy

The broad qualitative effects of agricultural deregulation can be forecast with some confidence. If nonagricultural policies remained unchanged during the transition phase, the main long-term effects would be as follows:

Production, land use, and farm structure

- Production, especially of grain and milk, would contract until excess supply all but disappeared; for important products such as grain, self-sufficiency would be less than 100 percent in normal years
- At least 10 percent and possibly as much as 25 percent of arable land would be withdrawn from crop production, most becoming pasture or forest
- The decline in farm numbers would accelerate
- The structural tendency toward more large full-time and small part-time farms, with fewer middle-sized family farms, would be intensified.

Prices, income, and capital

- Farmgate prices would be lower, more variable, and less uniform across producers and regions
- Farm incomes would decline, and asset values would fall drastically
- Gross agricultural investment would contract sharply; net investment might be negative for several years
- Retail food prices and national price inflation would be lower.[13]

Three important qualifications are in order. First, with Swedish economic policies in flux, further changes in food policy are possible in the near future. The nonsocialist coalition government formed in September 1991 is committed in principle to the 1990 food policy, but one of its first actions was to appoint yet another commission to "consider effects and factors that were not known or anticipated at the time of the [1990] decision, primarily the consequences of entering the EC" (MOA 1991a). Indeed, it appeared for a while in 1992 as if the Center and Moderate parties, together with LRF, would succeed in extending price supports, export subsidies, and the acreage set-aside—in the name of maintaining consistency with the CAP (Höök 1992a, Jägerhorn 1992b).

Second, one of the new policy's strengths—its built-in feedback and evaluation procedures—increases the likelihood that defects uncovered at the scheduled checkpoints will prompt corrective actions.[14] Here we

[13] If the rate of food-price increase slowed, labor unions' wage demands and general wage-price inflation would be slightly lower.

[14] The newly formed Swedish Board of Agriculture (SBA) will make annual progress reports and present its overall evaluation in 1994. One of its responsibilities is to signal a warning if the loss of farmland and other resources reaches a point where food security

Figure 8.1. Major casual pathways in the new food policy

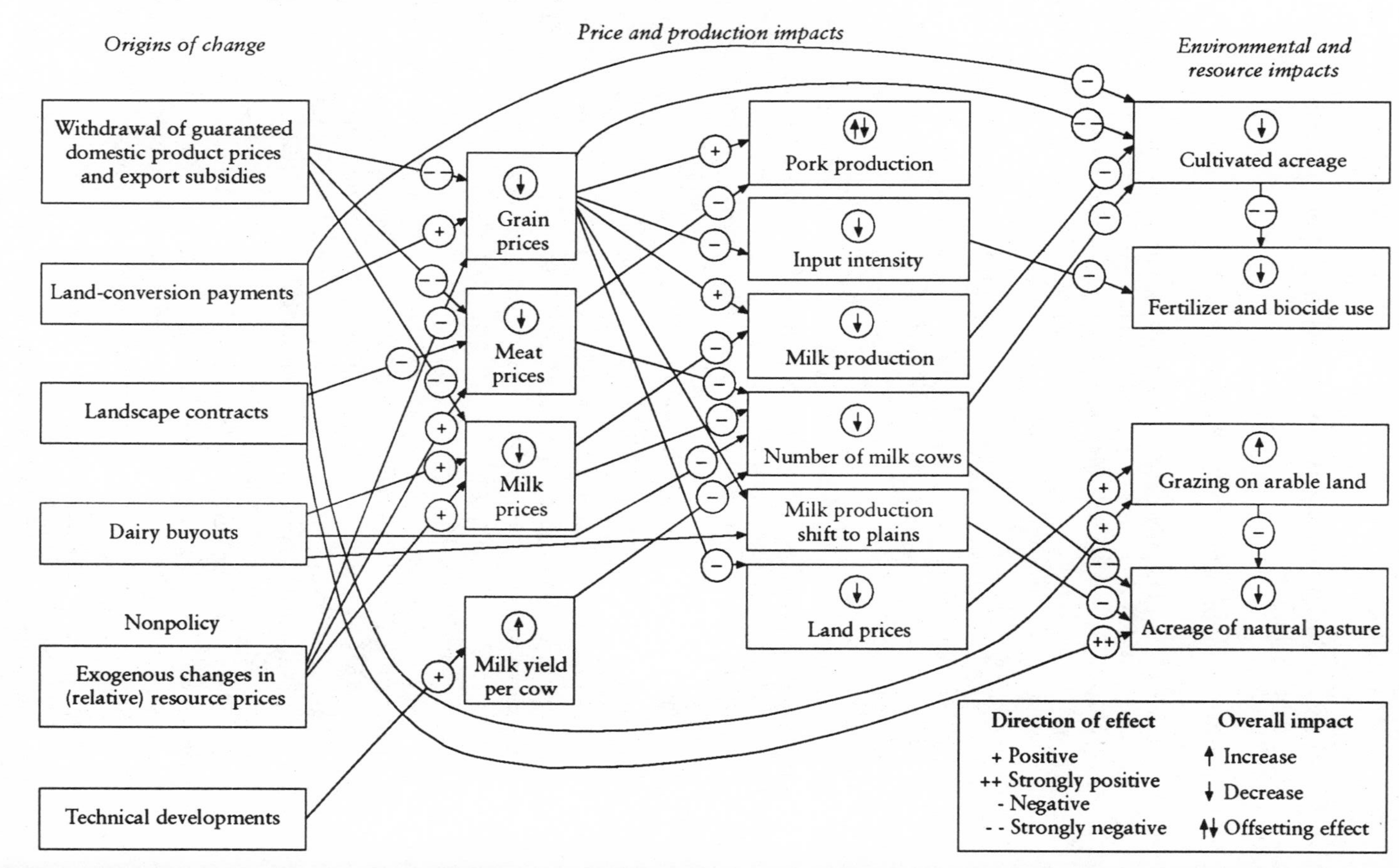

limit the discussion to likely consequences if the 1990 reform were fully implemented.

Finally, it is impossible to anticipate all the policy's indirect effects and subtle interactions between price deregulation and the policy's other measures. It is still more difficult to project the impacts of new regional, environmental, and tax legislation (explored in Chapter 9).

This prognosis relies largely on our own analysis, guided by a variety of governmental, journalistic, and scholarly sources. We lean most heavily on economic studies by Valter Johansson (1992) and Lars Jonasson (1991b). Figure 8.1 lays out the policy's main causal pathways; readers may find it a helpful reference as they make their way through the narrative.

Production and Marketing

Market deregulation is being introduced in a context of excess production: roughly 20 percent for grains (plus the capacity of set-aside land), 15 to 20 percent for milk and 10 percent for pork. Supply and demand for beef, poultry, and eggs were more or less in balance in 1990 (though there were often egg surpluses), and there were net imports of lamb and cheese. An analysis of consumer responses to price changes indicates that demand elasticity for basic foods is higher than is often assumed, so price deregulation is expected both to expand and to modify the consumer's market basket.[15]

With full deregulation, there are three possible market scenarios, depending on whether Sweden is a net importer, is self-sufficient, or produces an exportable surplus. The alternatives are illustrated by Figure 8.2 for a commodity such as wheat, whose supply fluctuates with climatic conditions. At S1 production is less than demand, so there will be some imports even at the tariff-protected domestic price, W+T. If supply is S_2, domestic production equals demand and the market clears at price P2. With a bumper harvest, S3, there is excess supply even at the world market price, W, so it will prevail in the home markets. Exports will occur. Since year-to-year grain yields typically vary by 10 to 15 percent, prices may fluctuate by as much as the

is threatened. The policy's environmental performance will be monitored and evaluated by NEPB.

[15] Silvander and Drake's (1990) own estimates and review of other studies suggest that demand may be price elastic for beef and pork but is inelastic for grains and milk. Own-price elasticities fall into the following ranges: pork, 0.31 to 1.31; beef, 0.21 to 2.41; fluid milk, 0.08 to 0.82; and cereals, 0.38 to 1.09. The cross-elasticity between beef and pork, relevant to a prediction made in the text, is roughly 0.3.

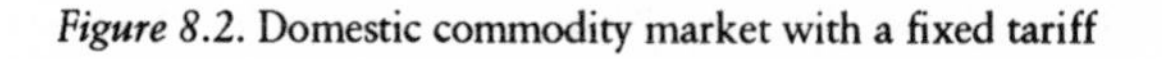

Figure 8.2. Domestic commodity market with a fixed tariff

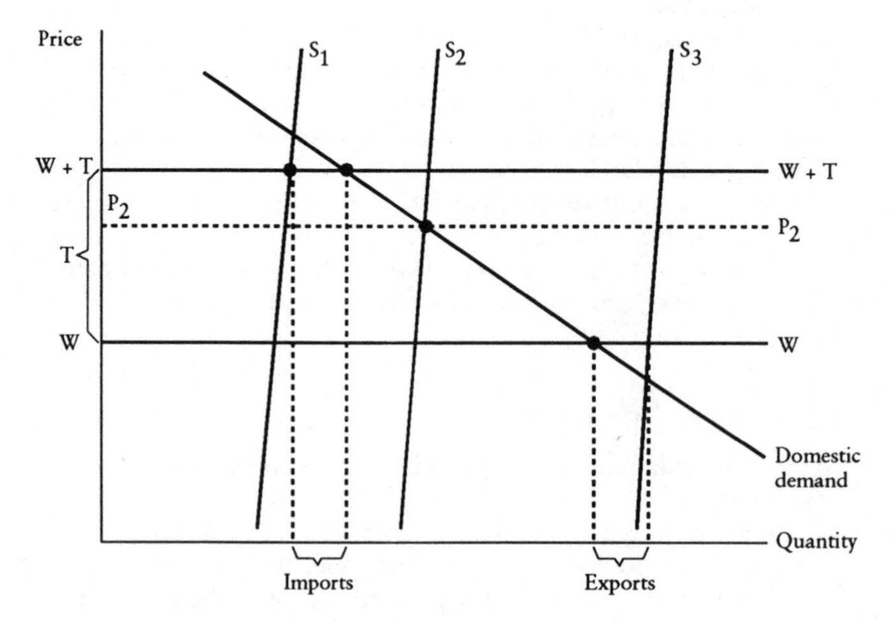

Notes: Shipping and transaction costs are omitted.
S_1 = short-run supply with poor harvest
S_2 = short-run supply with normal harvest
S_3 = short-run with bumper harvest
P_2 = domestic price with nonbinding tariff
T = tariff levy
W = world market price
W+T = domestic price with binding tariff

tariff, T. During the transition phase, and later on in bumper harvest years, surpluses of several commodities may persist, depressing prices below the long-term market clearing level. If a GATT accord were to impose tariff ceilings, however, Sweden would probably be a net importer of nearly all staple farm commodities in the long run (SBA 1991b).[16]

Over the thirty-month period from January 1989 to June 1991, the actual farm price index declined at an annual rate of 3.3 percent. Retail food-price inflation dropped from an average of 5.9 percent in 1986–90 to 2.3 percent in 1991 and *minus* 4.5 percent through August 1992. (As a point of reference, the general consumer price index was

[16] Since the majority of producers cannot cover their long-run unit cost at a price of W, either competition or collusion will ultimately force supply below S3.

still rising by 9.6%/yr in 1991.) Price deregulation and conversion subsidies get only partial credit for the lower inflation, however. Other factors, especially a cut in the value-added tax on food and antitrust actions against food distributors, also had major impacts (Nilson 1992; SBA 1991a,b,c).

On the whole, elimination of excess capacity is proceeding quickly, as farmers respond to the lower prices and conversion subsidies and as they act on their expectations about price and cost conditions after the transition. The initial cutbacks in grain acreage and dairy herds were larger than anticipated, evidently because the lump-sum payments for land conversion and dairy herd retirement are so much larger in the transition's first year. In 1990/91, 23,000 farmers applied for conversion supports covering 400,000 hectares; 350,000 hectares (14% of arable land) were actually enrolled. The enrollment figures for 1991 and 1992 were only 22,000 and 15,000 hectares, and the total of 387,000 hectares over three years was close to the predicted magnitude. In the first eighteen months of the transition, 3,500 dairy operations, more than 10 percent of the total, shut down (SBA 1992a). An allocation problem, to which we return in evaluating the policy, is that a sizable fraction of the converted acreage is fertile plains land rather than economically marginal land.

Food and Feed Grains. Many Swedish grain growers cannot compete in world markets, especially at prices depressed by widespread surplus dumping. So as the internal support price drops from US$0.22 per kilogram in 1990 to $0.15 in 1993, analysts expect one-third or more of grain land (at least 320,000 hectares) to shift out of grain production.[17] These calculations incorporate the effects of a small increase in direct cereals consumption and the positive demand effect of lower feed grain prices, but the contraction in dairy herds more than offsets these demand stimuli (Johansson 1992; Jonasson 1989, 1991b). Depressed land prices, together with conversion supports, will make forage crops, energy crops, and grazing profitable on some of the land taken out of grain.

During the transition, excess supply is likely to keep grain prices down near the minimum support level.[18] But it is predicted that the long-run domestic market clearing price will be about $0.18/kg, or 20

[17] Jonasson's (1992) upper-bound estimate is that 700,000 hectares, or nearly half of all grain and oilseed acreage, will be withdrawn from production in the long run.

[18] In 1991, ideal growing conditions led to a bumper harvest and kept prices down; in 1992, however, a severe drought in the south depressed yields.

Table 8.3. Impact of deregulation on livestock numbers

Animal	1990 numbers (1,000s)	Percentage change[a]
Dairy cows	576	−17
Beef cattle	75	+66
Calves	1,067	−8
Swine	2,264	No significant change
Laying hens	6,392	No significant change
Chickens	4,919	No significant change
Sheep (excl. lambs)	162	No significant change

Sources: Jonasson 1989, SS 1991b.
[a]The effects of transitional measures (land-conversion payments) and animal-protection regulations are not incorporated in these estimates.

percent higher than the final year's guaranteed price of $0.15. The price that actually prevails will depend on the level of tariff protection (SBA 1991a).

Livestock Production. The animal subsectors are affected by deregulation in complex ways (Figure 8.1). Since 70 percent of agricultural land has been used for fodder production, the derived demand for livestock feed is critical for long-term land use. Reciprocally, the cost of feed grain and forages has a bearing on the economic viability of livestock production. Lower grain prices and subsidized conversion of arable land to pasture, in particular, have a stimulative effect. There are no reliable estimates of the long-term changes in livestock numbers, but Table 8.3 summarizes Jonasson's simulation of likely magnitudes under full deregulation and all economic adjustments.

In pork production, a feed grain price at the predicted long-run level of US$0.18 per kilogram would cut unit costs by about 5 percent. If this cost saving were passed on to consumers, because of supply inelasticity, consumption might increase as much as 7 percent. Since this roughly equals the current excess supply, only a minimal cutback in pork production would be necessary (Jonasson 1989, Silvander and Drake 1990).

In the dairy sector, lower fodder costs will be swamped by the effect of lower milk prices. Jonasson (1989) expects a reduction of nearly 20 percent in milk cows within three years due to the current surplus of processed milk products, stagnant demand, and yield-increasing technological developments. The first year of the transition seemed to bear out this prediction, as two thousand milk producers (8% of the total) accepted payments to terminate operations, and nine hundred older farmers opted for early pensions. Together they accounted for 12.5 percent of fluid milk production. Actual milk production declined by 9 percent between 1990 and 1992. Depending on the pace of technical innovation and the stimulus of lower feed costs, excess milk could disappear by the mid-1990s (MOA 1993, SBA 1991c).

The elimination of measures to equalize milk prices means that farmgate prices will come to reflect differential transport and handling costs. This will stimulate more milk production near processing facilities and large population centers and cause a sharp contraction in other regions. Larger farms will have scale advantages in handling and transportation. The main exception to these generalizations is the Norrland, where continued price supports (motivated by regional and national security goals) will sustain production levels and small-scale operations. These farms, however, are likely to reduce their high-cost forage crop production and "import" more cheap grain from the south. (The impact on the northern landscape is discussed in section 8.4.)

Beef and veal production face cross-cutting influences. It was expected that liquidation of dairy herds would cause a large excess supply of beef and veal during the transition, so temporary export subsidies were legislated to offset downward price pressure.[19] Hundreds of small, geographically dispersed dairy farms are expected to shift into extensive part-time production of beef, veal, and replacement heifers, at least as long as the present generation of operators remains active. Indeed, while the number of milk cows decreased by 50,500 (9%) between 1990 and 1992, the total number of cattle, including heifer and veal calves and steers, increased by 57,000 (3%) (SS 1992:144). In the longer run, however, beef production will probably become more concentrated on large dairy farms and specialized production units, as hundreds more nonviable dairy operations close down.

Production costs for meat animals will fall as feed grain prices and rents on forage cropland decline. If, as predicted, grain prices fall relative to the cost of growing coarse fodder, feed rations will shift toward grain. Intensive confinement operations will be advantaged over more extensive grazing and forage crop systems. But a smaller effect in the opposite direction will be induced by subsidized conversion of arable land to permanent pasture and management contracts

[19] In fact, nominal beef prices were stable in 1991 without export subsidies, as off-take did not increase enough to depress prices much. There was an increase in the inventory of cows and calves kept as beef animals.

for nonarable grazing land. For several years at least, these measures will support extensive beef/veal operations and sheep grazing on several hundred thousand hectares. Sheep numbers grew by 100 between 1990 and 1992 (SS 1992:157).

Supply adjustment has been minimal for pork, as the market came into balance in 1991 with no significant decline in stocks. Changes have been slower than expected for beef, as the anticipated glut failed to materialize and downward price pressure was minimal. Until it ends in 1994, the beef export subsidy acts to raise the relative price of beef and induce a slight demand shift from beef to pork (SBA 1991b).

Other Commodities and Their Land-use Implications. The new policy's most conspicuous effect on land use is the afforestation of several hundred thousand hectares. The pace of tree planting is accelerated by the three-year limit on applications for conversion payments, but it is slowed slightly by farmers' tendency to use their machinery to grow regulated crops as long as price supports continue.[20] For nearly a century, most Swedish farmer-foresters have planted even-age spruce and pine stands, for silvicultural and commercial reasons. The land-conversion program, however, offers a premium of up to US$2,300 per hectare for deciduous and mixed forest, for reasons of aesthetics and biological diversity. The budget to back up this incentive is small, and its impact is not yet clear (Kumm and Andersson 1991).

As a food-security measure, contracts are offered to oilseed producers at a price high enough to maintain the pre-reform acreage. Actual 1991 plantings fell below the target, so prices have been further increased.

During the legislative debate, there was much talk about large-scale conversion of redundant grain land to industrial crops (including biofuels) and about profitable organic farming opportunities. The technologies, marketing infrastructure, and price incentives to stimulate such developments are not currently in place and have not been planned in detail. Without a major commitment of public resources, a market-driven agriculture is unlikely to shift more than 3 percent of arable land (100,000 ha) to industrial crops in the 1990s.

Structural and Geographic Effects

Agriculture's long historical contraction will accelerate and its geographic center of gravity will shift even farther toward the southern

[20] A glut of used farm equipment weakens the incentive to liquidate crop-production machinery quickly.

plains and coastal regions. To be sure, heavily subsidized northern farms will remain, but they will have little aggregate economic significance. Change will be hastened, on the one hand by lower, less stable, and less uniform prices and on the other hand by transitional payments for land conversion and dairy herd sell-offs. Scale economies, differential milk pricing, and a cheapening of feed grain relative to forage crops will increase the concentration of capital and production in large units. The contraction of food processing industries and other farm support infrastructures will also be greatest in economically marginal regions where small farms predominate (SBA 1991a).

Several countervailing forces are also at work. By reducing output prices and land costs relative to the cost of wage labor, deregulation will slow the drift toward capitalist agriculture. Further, the trend toward larger, more industrialized beef and veal operations will be somewhat offset by lower land rents and by payments for land conversion and landscape preservation, all of which favor extensive cattle raising. Hundreds of small dairy farms have the skills, buildings, and equipment needed for beef and veal production, so short-term growth in small, part-time beef operations is likely.

On the whole, the new policy reinforces the pre-reform tendency toward structural polarization: increased concentration of capital and production in large-scale units; persistence of small, part-time farms with minor economic but great cultural and environmental importance; and a disappearing middle. The dynamics are, of course, more complicated. For example, the conversion payments for grain land have attracted a disproportionate share of large farms in fertile regions. In 1990/91, farms larger than one hundred hectares (23% of total arable land) accounted for 33 percent of conversion hectares, while farms below twenty hectares (19% of arable land) had only 10 percent of the converted land. One obvious reason is that large commercial farms have proportionally more grain land eligible for conversion payments; but it also seems that numerous big operators have adopted a strategy of banking their conversion payments, "hibernating" a few years until Sweden joins the EC, and then returning to grain production. At the other end of the spectrum, most small farms have nonfarm income sources, making them less responsive to economic signals. It may also be more profitable for large capitalists to convert land and lay off wage earners, since family farmers have few alternative uses for their labor power in the short run.

We note, finally, that the demand for lease land has contracted, driving down rents and making the future of tendency relations unclear. The expected decline in fixed farm investment and input purchases has occurred. These changes, plus the elimination of dairy processers, will undermine the profitability of many businesses serving agriculture and weaken the economic fabric of rural areas. In some locales, support services will fall below the critical mass needed to sustain a farm economy (SBA 1991c, SS 1991b).

Welfare, Distribution, and Further Land-use Effects

The policy-induced contraction in input use and the accelerated shift of labor and capital from farming to higher productivity uses implies a small net economic welfare gain to society. The loss of consumer welfare under the old policy, roughly US$2.0–2.6 billion/year, gives a rough idea of consumers' potential gain from deregulation—if import protection for farm products and processed foods were eliminated and international commodity prices remained at their low 1980s levels. The net social welfare gain from deregulation is smaller, however, since import levies are still in place and the benefits to consumers are partially offset by lower farm incomes and the frictional costs of resource reallocation (Fahlbeck 1989).[21] A critical unknown is whether a GATT agreement will end export dumping and push up world prices.[22]

Potential welfare gains are vitiated by several aspects of the new policy. Most important, the continuing lack of tariff uniformity limits allocative efficiency gains from domestic market deregulation. Other things being equal, resources will flow into production of commodities with relatively high tariff rates and away from those with lower tariffs, distorting the linkage between supply and demand. Producers of a commodity whose supply meets domestic demand slightly above the world market price (e.g., grain in southern Sweden) receive little or no tariff stimulus (Figure 8.2). Meanwhile, farmers producing commodities whose costs are far above the international level (where Sweden has a comparative disadvantage) will benefit from the full effect of the import levy. In addition, the Swedish Board of Agriculture has concluded that deregulation will lessen the country's preparedness for

[21] Under Fahlbeck's assumptions, net social welfare gain would be roughly equal to the loss under the old policy, or about $115 million/year. This is roughly 0.07 percent of GDP.

[22] Calculation of welfare gains and losses requires knowledge of socially optimum prices. In the case of farm commodities, there is no straightforward way to estimate them, since nations intervene in many ways to distort prices and since important negative environmental externalities are not factored in as costs. Optimal prices would certainly be well above recent international levels; for some commodities, they might even be higher than under the old Swedish policy (see Blandford et al. 1990).

a prolonged international blockade, so the risk of a future welfare loss is increased by the new policy (SBA 1991a).

Deregulation will adversely affect present landowners, especially those in the fertile plains areas where grain-support prices were capitalized at the highest rates. It seems inevitable that those who benefitted most from the old policy will lose most from deregulation, even with transitional income payments and land-conversion subsidies to soften the blow. In the fertile southern plains areas, the implicit land rent may fall by 65 percent or US$280/ha./year and market value by $5,600/ha. This decline translates into a capital loss of $560,000 to the owner of one hundred hectares of prime land. The nationwide decline averages 80 percent: a loss of $130/ha in annual rent, $2,600/ha in market value, and $130,000 in total asset devaluation for a typical fifty-hectare farm.[23] Three years of income supplements will offset only about 3 percent of the expected capital loss (Jonasson 1992 and authors' calculations). Tenant farmers and new entrants will not be disadvantaged.

Once land values have depreciated and capital losses have been written off, the profitability of crop production in the most fertile plains regions should recover to nearly its pre-reform level. Under plausible assumptions about tariffs and demand elasticities for beef, its profitability may well increase with lower feed costs.

In much of the mixed and forest area, production costs on most land will exceed the value of output, even at a zero land rent. If this land is not protected by landscape-management agreements, most of it will sooner or later be abandoned or afforested. Part-time and hobby farming are scheduled to receive a small amount of investment assistance as part of regional development policy; this policy will help to keep some patches of land open in these regions. Niche product and value-added opportunities may also increase, but it is likely that ley and extensive beef production (along with heavily subsidized dairy operations in the north) will be the main surviving forms of agriculture (Jonasson 1992).

In sum, Swedish citizens will receive permanent benefits from the new policy as consumers, but they will pay somewhat higher taxes for a period of five years. In the first year of the transition, excellent grain yields combined with the various income compensation and conversion payment schemes to increase net farm income. Nonetheless, farm asset

[23] These figures are long-run equilibrium values, after completion of the transition period. Asset losses are derived from the lower rents, capitalized at a real interest rate of 5 percent.

owners, as well as owners and workers in farm-related businesses, will be the main economic losers. The aggregate economic gains outweigh the losses, but the combined income and capital losses of a typical farm family are not fully compensated and are far greater than the typical Swedish household's benefits (SBA 1991a, b).

Implications of Regional Changes for Food Security

In the past, food security has been interpreted as maintaining sufficient production capability in every part of Sweden, on the premise that hostilities might disrupt internal supply links. Although the greatest absolute income and capital losses from deregulation will hit large-scale farmers in the plains, the largest proportional contraction in farm activity and open land will be in the mixed and forest regions (apart from the Norrland). The principal direct cause is low milk prices, but farms will also be jeopardized by cheap feed grain from the south and by the demise of nearby dairy processors and input suppliers. The pool of farm skills and equipment will dwindle, and afforested land will not be readily reconvertible in an emergency. Large areas of the country will become incapable of meeting more than a minimal fraction of their subsistence needs in any unanticipated emergency (SBA 1991a).

Economic Appraisal

In the European setting, Sweden's new agricultural policy emphasizing consumer welfare as an end and competitive markets as a means is unique. Market forces, with all their uncertainties and potential for "creative destruction," are intended to play a much greater role in the future evolution of Swedish agriculture. The official rhetoric puts the consumer in the center, but it also elevates environmental protection to a primary policy objective, promising improvements on both economic and environmental fronts. Indeed, the new food policy *is* something of a magic bullet, since its permanent measues will bring commodity supplies into line with demand, stimulate more efficient food production, enhance consumer well-being, reduce (long-term) tax burdens, and cut pollution (see section 8.4). Nonetheless, it should be kept in mind that the transition phase has some perverse allocational effects and a high price tag for taxpayers. Furthermore, the magnitude of consumer benefits and efficiency gains hinges on two unresolved policy issues: import tariff levels and measures to increase competition in the food distribution system.

Some policy components were adopted with too little prior technical assessment and political consideration of detrimental effects and cost-effectiveness. A few of the most serious deficiencies have already been mentioned. Tariff distortions vitiate allocational and welfare gains. Price supports to sustain Norrland agriculture are not cost-effective compared with direct income support and targeted compensation for supplying public goods.[24] And food-security measures lack a comprehensive strategy for risk, especially the risks associated with what are essentially irreversible changes, such as afforestation and the exodus of farmers. There is also increased risk linked to external shocks, such as an international food shortage extending over several years.

Measures to smooth the transition to a deregulated farm economy and to minimize distributional inequities deserve special comment. On the positive side, the ethical and economic principles behind adjustment assistance are well founded. It is equitable to compensate economic actors for income and asset losses caused by unanticipated policy changes, especially since the old policy promoted their investments. It is rational to reduce frictions (time lags and transactions costs) when producers shift from one mode of agricultural production to another. And there is a public interest in speeding the flow of labor and capital to more productive uses. But negative effects of the specific transition measures for resource allocation, consumer well-being, and the environment could be avoided.

The central example, already cited, is land-conversion payments; at over $1 billion, they are the largest transitional cost. The enrollment criteria do little to ensure that the parcels of land taken out of crop production have the lowest opportunity cost (are the least fertile), that decisions about the land's new use consider its biological and aesthetic integrity, or that conversion will not undermine regional economic vitality. Thus, we observe withdrawal of some of the best grain land from cultivation and subsidies for planting trees and energy crops whose social benefits have not been persuasively demonstrated. Since open farmland is both a renewable resource and a source of public goods, one-sided financial incentives to cut back this input are questionable (Drake 1992).

[24] Artificial milk and meat pricing continues to be the main form of Norrland support, despite the LAG's strong recommendation in 1989 that these measures be replaced by direct income support. Dissatisfaction with the new policy's regional and environmental effects and the likely need to conform with EC measures for disadvantaged rural areas led the government to create yet another parliamentary commission on Norrland agriculture. It recommended ways to strengthen the public goods benefits, reduce the bias toward economi-

From a macroeconomic perspective, the state's budget constraint means that dairy herd buyouts and land conversion payments must be financed from higher taxes. Since Sweden's tax rates are already exceptionally high, any further increase is likely to have adverse incentive effects on some other economic activity. All in all, it can be argued that the transitional reallocation of agricultural resources (as distinct from income compensation and payment for public goods) would have been better left to market forces.

The transitional measures are also a dubious way to put the consumer at the center. By accelerating supply contraction, they tend to keep short-run farm prices close to ceiling levels (the world price plus tariff, in Figure 8.2).

Finally, incentives to accelerate land conversion and dairy herd selloffs are questionable in light of Sweden's EC application. It is doubtful whether it makes either political or economic sense to speed agricultural contraction before joining the EC. Since the Common Agricultural Policy uses revenues collected from all members to compensate farmers and finance structural grants to member nations, Sweden seems to be preemptively shrinking its own future slice of the EC pie. More ominous, the former grain producers who have been paid to take highly productive land out of production, planning to hibernate until Sweden enters the EC, may be in for a rude awakening if the CAP forbids idled land from being returned to production. (These questions are explored more fully in Chapter 9.)

In summary, agricultural deregulation will benefit Swedish consumers and contribute marginally to the economy's dynamic efficiency. The new permanent policy instruments are based on sound principles, though they are not perfectly implemented. Our central economic criticism is leveled at the transitional measures, especially the land conversion payments and dairy herd buyouts, which are fundamentally flawed.

8.4 Pale Green Environmental Effects

The new food policy has goals of "minimizing the negative environmental effects of agriculture" and maintaining its provision of public goods, using directly targeted instruments. The policy's major environmental impacts are:

---

cally advantaged locations, discourage input intensity, and raise food processing efficiency (MOA 1991b).

Table 8.4. Expected changes in arable land use (1,000s of hectares)

Crop	1990 actual	1990 simulated, deregulated	2000 simulated	
			Regulated	Deregulated
Bread grains	423	257	400	291
Feed grain	912	606	886	657
Oilseeds	168	99	170	120
Ley	778	683	529	545
Pasture	191	339	148	161
Other crops	117	166	156	146
Fallow	222	88	[a]	[a]
TOTAL	2,811	2,236	2,289	1,920

Sources: Johansson 1992 (2000 simulation), Jonasson 1989 (1990 simulation), SS 1991b.

[a]Fallow not included in modeling.

- Reduction of chemical and fertilizer inputs
- Reduction of aggregate energy use
- Contradictory influences in the agrarian landscape. Improved scenic and biological diversity in the southern plains will be offset by an overall loss of open land, especially highly valued meadows and natural pastures in the mixed and forest regions.

The direct effects of targeted measures, interacting with the indirect effects of market deregulation and transitional measures, are considered under the same headings used in Chapter 7.

Pollutants

Nitrogen leaching will be cut back in three ways: a lower optimum application rate, with lower crop prices; a lower proportion of cultivated land in high-leaching crops; and a decline in total crop acreage. By 1994, lower grain and oilseed prices are expected to reduce nitrogen input by about 10 percent (25,000 m) and leaching by about 5 percent (2,500 mt). Most of the 350,000 hectares receiving conversion support will be put into extensive pasture and forest, cutting leaching by another 2.5 to 5 percent. In the livestock sector, the opposing effects of fewer cattle and more grain in feed rations have not been calculated. (Estimates are imputed from NEPB 1990c and SBA 1991a.)

Biocide use is expected to be cut about 10 percent, or 200 mt, primarily because of acreage reductions for bread grain and sugar beets. These crops account for 80 percent of the predicted decline.

Greenhouse gas emissions will also be curtailed, but to an unknown extent. Carbon emissions will be affected by the decrease in cultivated acreage and the relative contraction of energy-intensive land uses, although on some cropland energy use may increase as tractor cultivation replaces herbicides. Bioenergy crops that continuously cycle carbon may replace a small amount of fossil fuel in the national energy mix (see Chapter 9). And smaller numbers of cattle will mean lower methane emissions.

Landscape Preservation and Nature Conservation

The new policy's effects on different agrarian landscapes are complex and contradictory. Forecasts summarized in Table 8.4 indicate that the package of deregulation and transitional measures will accelerate aggregate farmland loss and modify what remains of the domesticated landscape.

Of greatest importance, most of the economically submarginal land with high biological, cultural-historic, and recreational value is likely to be abandoned or afforested in the long run. Transitional measures encouraging dairy farmers to sell out or retire quickly could be especially devastating in the mixed and forest regions, and indeed in any area remote from dairy-processing plants. In some regions, as much as one-third of the remaining open land is expected to be lost. These negative effects are somewhat mitigated by permanent price supports to Norrland dairy farmers and targeted payments for landscape protection, as well as conversion payments for wetlands and deciduous tree planting.

The situation in the southern plains is sharply different from that in the marginal farming areas. In the south, deregulated markets and some of the conversion options (though not energy forest) give farmers incentives to convert their land from intensive grain and oilseed production to more extensive uses such as ley and pasture, which enhance the visual and biological richness of local landscapes.

Jonasson's (1989) comparative static modeling of long-run equilibrium land use, with and without market deregulation (the 1990 simulation in Table 8.4), forecasts a 500,000-hectare reduction in annual crops offset by almost 400,000 hectares more ley, pasture, and rotational fallow. Johansson's (1992) acreage forecast for the year 2000 assumes that there would be a 250,000-hectare decline in cropland in the 1990s even under the old policy. Deregulation per se is expected to reduce farmland by an additional 350,000 hectares over ten years, with a decline in every category except minor crops. These simulations confirm the preceding interpretation: the increased proportion of ley and pasture in the plains would bring environmental improvement there, but the loss of much of the remaining open land in the mixed and forest regions of south and central Sweden would be socially and environmentally detrimental.[25]

Roughly 14 percent of Sweden's arable land (387,000 hectares) has been enrolled in the conversion program. In the south and the Norrland, the figure is 10 percent, but in the mixed and forest zones of central Sweden, 20 percent of arable land will ultimately be converted under this program. Participants are not bound to a specific land use, and by 1992, only 73 percent of the enrolled land was covered by a conversion plan and only 46 percent had actually undergone conversion. A survey in that year suggests that in the near term, 45 percent of converted arable land will be used for extensive grazing while another 28 percent will be shifted to other approved uses. Until the current generation of operators retire, the core enterprise on most former dairy farms in disadvantaged regions will be part-time beef production, a land-extensive activity that also has positive environmental effects (SBA 1992a).

Forty-three thousand hectares have been approved for energy crops, 10,000 for energy forest, and 33,000 for long-rotation forests. Surprisingly few farmers indicate an intention to plant conifers, but this may just be a matter of time. Many farmers are awaiting government signals about price supports for energy forest and especially about the implications of EC membership (SBA 1992a). The three-year budget for reconstruction of wetlands and planting of deciduous or short-rotation energy forest will cover only about 40,000 hectares (1.7% of arable land) and enrollment applications in 1991 quickly exhausted the budget appropriation (SBA 1991c).

From 1993, landscape-management agreements and NOLA contracts for maintenance of meadows and natural pastures are budgeted at $55 million per year. This amount should cover more than 200,000 hectares. Officials claim that the law requiring advance notice of any land-use conversion will limit the risk that land with a high public goods value will be abandoned or afforested before the state can intervene to protect it. Past enforcement of this requirement has typically

[25] The government asserted in 1990 that lower crop prices would improve landscape values by weakening the incentive to remove such field elements as stone walls and tree islands. Hasund (1990a), however, shows that the principal motive for removing elements, to save on labor and machine costs, would not be much affected by lower crop prices.

been "too little, too late," and there are no extra funds to purchase such vulnerable lands; thus we find this a dubious proposition.

There are two principal reasons to expect that much of Sweden's biologically rich and scenically attractive meadows, natural pasture, and marginal grass ley will eventually be abandoned or afforested. First, funding for management agreements covers less than half this land, and second, the new policy enhances the comparative advantage of large-scale dairy farming in the warmer, more fertile south. This effect will undercut the economic viability of grazing on low-productivity permanent pastures. Conversion payments for grazing on arable land in the south and for dairy herd buyouts in mixed and forest regions have accelerated the exodus of small milk producers, who formerly maintained much of the most valuable pastureland (SP 1990). The biggest unanswered question is how much "dark spruce forest will invade the smiling cultural landscape" over the coming five to ten years.

Sustainable Resource Use

The new food policy will affect sustainability through three main pathways: the intensity of variable input use, cultural practices, and the long-term availability of land and human resources.

As mentioned, cutbacks in fertilizer and biocides per hectare and in aggregate will be induced by lower crop prices, lower land values, and reduced rowcrop acreage. Studies indicate that lower grain and land prices strengthen incentives to plant more green manures and add legume forages to rotations, further lessening nitrogen fertilizer needs and leaching (Molander 1992). On the other hand, the accelerated withdrawal of grain land via conversion payments might cause a slight short-run increase in input intensity on remaining cropland.

Financial support for conversion to organic farming has been terminated, since its prime motivation, limiting excess grain and milk supplies, is expected to disappear. Transitional land-conversion payments may be used to support chemical-free niche-product enterprises on former grain land, but there is no evidence that this is happening on a significant scale.

Sustainability and food security are connected. Land shifted from grain to ley or pasture is used more sustainably in the short run and would still be available for intensive cultivation in a food emergency. Conversion subsidies for energy or long-rotation forest, in contrast, work against food-security goals, as does the large-scale loss of open land in disadvantaged regions. The stock of skilled farm operators in these regions will shrink along with the land base. The importance of accelerating these trends, as noted, depends on one's assessment of the risks connected to food-scarcity scenarios.

The development of renewable biofuels from agriculture is likely to be a key sustainability issue in the 1990s. A particular type of national security—energy security—prompted research and development on agrobioenergy back in the 1970s. Thus far, conversion grants have been made for planting 43,000 hectares of energy crops and energy forest. Biofuel advocates, including the Center and Liberal parties, the LRF, and some environmental organizations, also propose long-term price supports for bioenergy crops on the grounds that more bioenergy means less dependence on imported nonrenewable fuels and fewer greenhouse gas emissions. A big push for renewable energy would keep some land, labor, and capital resources in agriculture while maintaining farm incomes and property values, particularly in central Sweden. But the environmental impact and economic cost-effectiveness are problematic. We consider them more fully in Chapter 9, in the context of Sweden's overall energy strategy.

Animal Health

The new food policy did not address the goal of improved well-being for farm animals, and it has only minor and indirect effects in this direction. Expansion of extensive cattle production, with increased grazing and more coarse fodder in rations, should be beneficial. But for livestock in confinement operations, including pigs and poultry as well as cows and steers in large dairy/beef units, no significant improvement is likely.

A Green Assessment

The new food policy's green principles are laudable: it seeks to minimize agriculture's negative environmental effects and strengthen its contribution of collective environmental goods. The policy reflects a clear understanding that, in the coming market-directed regime, the state must mediate market forces to internalize negative externalities and stimulate the supply of public goods. There is a clear recognition that directly targeted instruments are needed to attain these goals efficiently. The new policy improves on the old in this respect, but in practice its mix of permanent and transitional measures fails to internalize externalities consistently or with an adequate commitment of resources.

It seems to us that environmental protection is less a core goal than a consideration or side condition applied to measures aimed at other ends. In the legislation, environmental goals were given much less operational content than economic goals. Whereas sustainable agriculture implies a multigenerational commitment, the new food policy does not spring from a transcendent vision or any first principles of long-term sustainability.

On the positive side, deregulation will reduce agricultural input intensity and pollution in general, while improving biological diversity and landscape amenities in the southern plains. In this limited sense, the policy's permanent measures *are* a magic bullet with both economic and environmental benefits. The targeted measures for wetland reconstruction and deciduous tree planting will also have beneficial effects. Norrland supports will keep some dairy farming and biologically rich landscapes alive in that region. And long-term management contracts will protect some of Sweden's arable land and natural pastures.

On the downside, the US$55 million annual budget for landscape-management agreements is far less than citizens would willingly pay for this public good. It is much too little to attain the policy's goal of "safeguarding a rich and varied agricultural landscape," and cumulative economic pressures are likely to overwhelm this small financial commitment. Since the 1930s, almost three-fourths of Strindberg's treasured *hagmark* has been lost, and with time the greater part of what remains is likely to be abandoned or afforested. This is perhaps the most serious green deficiency of the new policy, but other weaknesses can also be pointed out.

Keeping milk price supports as the central method of maintaining Norrland agriculture is not only economically cost-ineffective but also creates a bias toward greater fertilizer use (MOA 1991b). In most settings, land-conversion payments for energy forest or long-rotation forest have two-sided environmental effects and are debatable economic investments. Market forces will eventually eliminate nearly all agriculture in the mixed and forest regions of south and central Sweden. From a risk-averse perspective on sustainability, this trend—and much of the subsidized tree planting on arable land—can be challenged. One need not subscribe to neo-Malthusian predictions of long-term food scarcity to believe that Sweden is abandoning too much agricultural capacity—often the wrong capacity—too hastily. The Swedish strategic response to external risks in "the new world order" needs more analysis and public debate. Finally, the transitional measures regarding land use and green input taxes were not fully coordinated with the array of green measures initiated in the 1980s (discussed in Chapter 7).

8.5 Room for Improvement

The 1990 food policy was the end result of a characteristic Swedish exercise in compromise and consensus building. It is a reform that responds to manifest defects in past policy, not a transformation guided by a shared future vision. As participant observers and constructive critics, we suggest here how the arts of policy design might have been used to achieve the new policy's stated goals more fully and at lower cost.

The core problem, inconsistency between the policy's means and ends, has two aspects. First, the transition measures violate the basic principle that market forces should guide agricultural resource allocation. Second, the principle that beneficial and detrimental externalities should be interalized via directly targeted measures is only partially and inconsistently applied in practice.[26]

The Market Principle

The legislation states clearly that consumer demand and competitively determined production costs should guide farm resource allocation decisions, except where externalities justify state intervention. The transitional measures violate this principle, especially the land-use distortions induced by conversion payments. In general, those payments create a one-sided bias toward withdrawing land from production. Since farmland is a renewable resource with few alternative uses (a low opportunity cost), this bias does not make economic sense. Why not target transitional measures to accelerate the reduction of fossil fuel and fertilizer inputs? Measures should at least encourage retirement of land parcels with lowest opportunity cost, but in practice the policy has financed conversion of fertile fields in the southern

[26] It appears that Swedish citizens are beginning to appreciate the importance of means-ends consistency. A 1992 opinion survey revealed that 64 percent of Swedes believe that agricultural landscape preservation should be financed through taxes; only 20 percent view higher food prices as an appropriate mechanism. Twelve percent believe that no compensation is justified (AERI 1992:27).

plains. Conversion payments are being made for new land uses—forest and energy crops—whose social benefit relative to cost has not been persuasively demonstrated. In sum, the policy would be more cost-effective if compensation for farmers' losses were separated from incentives that distort resource allocation.

Paying for Public Goods: The Centrality of Side Effects

For many years, Sweden has officially sanctioned the Polluter Pays Principle (PPP) in dealing with detrimental externalities. Appropriately, the new food policy adopts what we have labeled the Producer Compensation Principle (PCP) for beneficial externalities.

It has been shown that PPP was frequently violated under the previous policy, as when livestock farmers were given public grants to help finance manure storage facilities. Now the monumental challenge is to classify, quantify, and reward the provision of agricultural public goods. Furthermore, as the farm sector ceases to have its own distinctive policy, it is imperative that general economic and environmental policies take agricultural side effects into account. We have seen that dairy farms in marginal farming regions contribute several types of collective goods. Yet a host of recent policy changes, including broad tax and antitrust laws as well as the new food policy, have driven several thousand of these farms out of existence in short order. The economic policymaking process should routinely include assessment of such effects and formulation of countermeasures.

With the eclipse of food security as a justification for agriculture's privileged policy treatment, other collective purposes have become more prominent. Since most of them depend in one way or another on the maintenance of farmland, we emphasize this issue. The food policy explicitly calls for protection of a "rich and varied landscape," and $55 million/year has been allocated to this purpose. But the present landscape-management contracts fail to fulfill the PCP in two distinct ways.

First, $55 million/year is far too small an allocation. Survey research indicates that citizens' willingness to pay to protect Sweden's present farm landscape and avoid extensive spruce planting on arable land is on the order of $200–600 million/year (at 1986 prices) (Drake 1987).[27]

[27] The notion of willingness to pay to preserve a resource's nonmarket values, as estimated by the contingent valuation method, must be qualified. Recent public opinion polls and voting behavior indicate that a large proportion of Swedish citizens are not willing to pay more total taxes to finance a larger public sector. Therefore, a larger budget for farmland

Second, virtually all farmland, even the south's homogeneous grain-oilseed landscape, produces some noncommodity benefits along with its yield of marketed output. In principle, then, all farmland should receive a rent from the public treasury, differentiated according to the land's public goods contribution.

Multiyear management agreements may be an effective administrative tool to protect a few hundred land parcels with exceptional collective values. But the administrative cost of applying this particular method to Sweden's nearly one hundred thousand farm holdings and 2.8 million hectares would be prohibitive. We recommend a differentiated annual rental payment to all farmland, like the one proposed in the 1990 debate by the Swedish Nature Conservation Society. This would entail a straightforward administrative mechanism and a simplified version of the land-rating system currently used for management contracts. This system weighs biological, recreational, aesthetic, and cultural-historic values. An initial land survey followed by random annual checkups would be necessary. With such a system in place, the owner's land-use decisions would depend on the land's combined rent from commercial production and public goods provision. An even simpler administrative solution would be to subdivide the nation into a small set of regions and establish rental payments for, say, three to five types in each region.[28] Beyond these techniques for enhancing consistency and cost-effectiveness, we stress several other ways to improve policy.

More Precise Goal Specification

Legislative objectives such as putting the consumer at the center and maintaining a rich and varied landscape are too general to provide much guidance in detailed policy design. Particularly in the case of environmental objectives, Sweden's food policy needs more precise goals to facilitate a rational choice of instruments. Lawmakers failed to establish a clear goal hierarchy within the food policy, so it is difficult to weigh various objectives, for instance when limited funds must be allocated to competing purposes (e.g., land-conversion payments

preservation would presumably require cuts in other budget items. Our survey research did not frame the question of willingness to pay in precisely these terms.

[28] These proposals do not rest on an assumption that all farmers are rent maximizers or that economic calculations alone would govern their decisions. Land use would, as always, be influenced by the owners' and operators' aesthetic sensibilities, moral convictions, and love of farming. Under the proposed system, an especially prized landscape could be protected by public purchase if economic incentives failed.

to pasture vs. forest) and when measures to enhance one objective will detract from another (e.g., manure storage requirements and the loss of small dairy farms). Consequently, administrators will retain considerable de facto influence over priorities via day-to-day decisions affecting policy implementation. The likely effect is greater inertial resistance to intended reforms.

A Firm Analytic Foundation for Policy Decisions

The choice of policy instruments often seems to be based on intuition and prejudice. Among the new food policy's transition measures, land-conversion payments are a prime example. At times, political leaders and bureaucrats rely on casual calculations even when systematic analysis is available. Indeed, such analysis is often contained in the very reports they have commissioned. Our point is not that technical experts set the political agenda, but that choosing measures without taking advantage of expert knowledge wastes research investments and often means higher costs and lower benefits than would be possible. The food policy's methods of paying for public goods are prime examples of missed opportunities to increase cost-effectiveness. We do not believe it is naive to hope that the quality of policy decisions could be improved by investing more to educate politicians about methods to evaluate the policy alternatives reviewed in all the studies they commission.

Risk Strategy

As we have indicated, some progress has been made in reshaping Swedish food-security measures to reduce vulnerability to a blockade of farm inputs. Biocide taxes are also being redesigned in accordance with risk profiles. These and many other agricultural policies could be made still more effective: first, if they were based on clearly stated and consistently applied principles for weighing various types of risk, and second, if the reports of expert commissions routinely included risk assessments of alternative policy measures.

Risk takes many forms, of course, ranging from straightforward pecuniary gains and losses to threats to human health, national security, biological diversity, and the global climate. It is a political choice whether to adopt a highly risk-averse strategy toward, say, a low-probability food-crisis scenario, or whether to adopt some other means of weighting—or valuing—possible outcomes. In the case of some potential environmental hazards, Sweden's public pronouncements stress the cautiousness principle: when in doubt, err on the side of avoiding collective risks. We would urge that this principle be applied more vigorously in practice. The point is that a more transparent valuation of different types of risk would facilitate both more useful technical analyses and more cost-effective selection of policy measures.

Politics is the art of the possible, and it would be unrealistic to expect or demand perfection in policy design. These suggestions for improvement are offered in a constructive and, we think, practical spirit. The critique should not obscure the impressive steps Sweden has taken to replace an obsolete agricultural policy regime with economic and environmental measures that better promote Swedish citizens' material interests and fulfill their green commitments. The new policy is not cast in stone, and as the action shifts from legislative hearing rooms out to the countryside, we are cautiously optimistic that proposals such as ours will be heard.

9

New Problems and Priorities in a Changing World

The 1990s are shaping up as a decade of major changes, both in Sweden's domestic economic and environmental policies and in the world order, where Sweden is a small player. Domestically, the strange brew of neoliberalism and greening is already modifying the food policy's impact. There have been a major tax reform, new antitrust regulations, privatization measures, a revision of energy strategy, and several environmental initiatives. Internationally, Sweden's new agricultural policy will be powerfully affected, if not stood on its head, by the end of the Cold War, GATT trade liberalization, and membership in the European Community. All these changes interact with the new food policy, necessitating revisions and shaping outcomes.

9.1 Agricultural Feedback from Shifting Domestic Policies

Nonfarm policies have always exerted a powerful influence over Swedish agriculture's immediate conditions and its long-term trajectory. Several important new nonfarm policies will interact with food policy—in complementary or conflicting ways—to determine aspects of Sweden's rural future ranging from water quality to employment opportunities and from scenic vistas to historic landmarks.

Tax Reform

A central component of the neoliberal shift is a fundamental overhaul of the tax system, begun in 1990. In recent decades, Sweden has had the highest aggregate tax level of any industrial nation. The core reform element is a sharp reduction in marginal income tax rates, with revenue losses offset by a broadening of the tax base (through elimination of many income shelters and expenditure deductions) and a value-added tax on energy. On the whole, the reform adversely affects the agricultural economy by removing favorable tax treatment of farm income, losses, and assets. This results from closing loopholes thought to be inequitable or to cause resource misallocation. The measures described below directly affect farm owners' after-tax income. Through their impact on farmers' spending power and incentives, they will indirectly affect the economic vitality of rural areas, landscape preservation, and maintenance of historically significant farm buildings.

The right of owner-operated enterprises to deduct business losses from other income for tax purposes has been eliminated. Previously, thirty thousand farmers, cultivating one-fifth of Swedish farmland, took advantage of loss deductions. Part-time and hobby farmers, mostly in the mixed and forest regions, will be particularly hard hit. The impairment of their economic conditions will sooner or later have a negative effect on the maintenance of open space and cultural-historic farm buildings, and on the general economic vitality of marginal farming areas.

The economics of part-time and hobby farming is also adversely affected by taxes that penalize rural commuters: tighter deductability restrictions on commuting costs and a higher carbon tax on fuels (discussed below).

Furthermore, maintenance, repair, and interest expenses on farm dwellings can no longer be deducted as business costs. This raises the average farm's tax liability by $335 per year, especially affecting owners of large, old farmhouses with high maintenance costs—precisely those that contribute most to public goods. On the other side of the ledger, lower inheritance taxes improve the next generation's prospects for keeping farms intact and in the family (Aronsson and Pettersson 1989, Hasund 1991).

Also on the positive side, removal of the silviculture tax on private woodlots in 1992 improves farms' cash flow, especially in mixed and forest regions where well over half of farm holdings are forested. This improves small farms' economic conditions but also slightly strengthens the incentive to plant trees on open land. Property taxes on undeveloped real estate have also been slightly reduced, which marginally raises farms' cash flow and weakens the incentive to sell land for development.

Finally, the value-added tax on food was selectively cut from the general 28 percent level to 18 percent in 1992 in response to public frustration over rising food prices.[1] (Such a selective tax cut contradicts the government's objective of creating consistent market signals.) Retail food prices will fall by roughly 5 percent, increasing demand for raw foodstuffs by a percent or two and marginally improving farm prices.

Antitrust

In the drive to purge inefficiencies and inequities caused by collusive market behavior, a commission was appointed in 1989 to investigate three particular sectors: food, transport, and housing. Its core recommendations for the food sector, subsequently adopted by the Riksdag, were to prohibit collusive price fixing, market sharing, and collectively financed export dumping. These actions were viewed as necessary to fulfill the promise of the food policy reform. The domestic food distribution industries will be less protected from processed food imports and from multinational food corporations. Agriculture will feel the effects in several ways. Lower retail food prices will stimulate a slight increase in demand for raw foodstuffs, but they will also put pressure on producer cooperatives to pay lower prices to farmers. Foreign-owned processors will presumably be less committed to buying locally than Swedish processors owned by farmer cooperatives.

Perhaps most important, the cooperatives' tradition of cross-subsidizing disadvantaged farmers by equalizing prices across farm sizes and locations will be much more difficult to maintain in competitive markets. Prices to small and remote farms will tend to be depressed, and when this disadvantage is coupled with higher tax bills, the survival prospects of small full-time farms will be jeopardized (MCSA 1990).

New Environmental Regulations

The Commission on Environmental Taxes was appointed in 1988 to assess the contribution of economic levers to more cost-effective environmental protection. Agriculture was one focus of its 1990 report. For the first time, the fertilizer tax was evaluated in terms of a

[1] The last of the retail food subsidies introduced in the 1970s was removed as part of the general tax reform.

tangible goal: a 20 percent reduction in total nitrogen applications by the year 2000. The commission recommended that the nitrogen tax should be increased by 40 percent (US$0.10/kg) to $0.35/kg in 1990 prices, when the nitrogen levy used to finance surplus grain exports is no longer needed. In practice, this recommendation has been stood on its head: the government has decided against a higher nitrogen tax, while at least temporarily raising—rather than eliminating—the levy used to finance exports. In 1992, the State Board of Agriculture went so far as to urge that the tax be abolished, since it has had so little effect on farmers' practices and since it will have to be removed when Sweden enters the EC. SBA analysts predict that the goal of reducing nitrogen applications by 20 percent will be attained without the tax, as a result of acreage contraction and lower fertilizer profitability (SBA 1992c).

The Commission on Environmental Taxes also proposed that biocides be taxed according to hectare doses instead of grams of active ingredient, since many high-potency chemicals are now used in such low volume that the effective tax is nearly zero. To reach the 1995 target of a 75 percent reduction in aggregate use and also induce a shift to less harmful compounds requires a dosage tax that reflects a chemical's risk, including its toxicity, biodegradability, and mobility in the soil and food chain. The Riksdag has imposed a uniform tax of $3.30 per hectare dose and appointed an expert committee to develop the risk-classification system for a differentiated tax (MOE 1990a, SBA 1992c).

The political parties have agreed that cadmium in phosphate fertilizer should be restricted to less than 5 milligrams per kilogram of phosphorus, but a target date and policy instruments have not been selected. The Commission on Environmental Taxes recommended a cadmium tax, with a corresponding reduction in the general phosphorus tax, but suggested that its introduction be delayed until 1993 to give superphosphate manufacturers time to install more effective cleaning technology. Following this advice, the government has submitted a bill that would impose a tax of US$5.00 per gram on the cadmium content of superphosphate fertilizers (GP 1993). An expert committee appointed to refine the instrument design has recommended a maximum tolerance level of 100 mg/kg of phosphorus, similar to the standard in several EC nations. In June 1993, a cadmium tax was also legislated (MOA 1992, MOE 1990a).

By order of the Ministry of Agriculture, the National Board of Agriculture is investigating the feasibility of a 50 percent reduction in agricultural ammonia emissions by the end of the 1990s. The combined deposition from domestic and foreign sources is much higher in the south than in the rest of the country, so NBA was asked to submit an action plan for a 25 percent cutback in the four southernmost counties by 1995. The plan recommends, among other things, that farmers in the region be required to fill liquid manure stores from the bottom to minimize denitrification, that manure stores have lids, and that manure be plowed down within four hours of spreading (vs. twenty-four hours in other regions). It is estimated that these measures together would reduce local agricultural emissions by 30 percent at a gross cost of $5 million per year.[2] As with the manure management regulations discussed in Chapter 7, compliance costs vary tremendously: from $0.20 to $10.00 per kilogram of abatement, depending on the technique employed and specific farm conditions (SBA 1991e). The package of recommendations would balance environmental and economic objectives fairly well, but it has avoidable flaws in enforcement, equity, and cost-effectiveness. The tilling-in requirement is difficult to enforce, the measures are biased against farmers who have already invested in liquid manure systems, and the command-and-control approach results in weak incentives for further technological development.

In the deregulated market environment that is taking shape, these regulations for managing animal wastes and those discussed in Chapter 7 will move Swedish agriculture closer to the Polluter Pays Principle and toward efficient pollution abatement. They will also raise Swedish production costs relative to countries with less stringent waste-management rules. Recall that costly animal welfare regulations also come into effect during the 1990s. If and when a GATT treaty liberalizes the livestock-products trade and Sweden joins the EC, these costs would disadvantage Swedish producers and encourage imports from less environmentally responsible nations. Such subsectors as pork, poultry, and eggs could be especially hard hit. Finally, since there are scale economies in manure handling, some marginal dairy operations would be forced out of business by the new requirements.

National Energy Strategy and Agricultural Bioenergy

National energy strategy is one of the most divisive issues facing Sweden's present nonsocialist government, since it affects international economic competitiveness, employment prospects, material living

[2] These measures conserve nutrients, so manure's nutrient value would be increased. If leaching and denitrification losses in stored livestock wastes manure were totally eliminated, for example, the net cost would be only $1 million per year (SBA 1991e).

standards, regional vitality, and national security—as well as environmental quality. In January 1991, shortly before it lost power, the Social Democratic party allied with the Liberal and Center parties to push through a new energy policy with the intention of resolving glaring goal conflicts. Existing energy legislation called for decommissioning all nuclear power plants by 2010, starting in the mid-1990s;[3] prohibiting hydroelectric development on the four major remaining wild rivers; limiting Swedish vulnerability to fossil fuel market shocks; and freezing carbon dioxide emissions at 1990 levels. As if these challenges were not great enough, it was also seen as imperative to maintain the international competitiveness of energy-intensive export industries and the employment base of northern and interior regions where they are concentrated.[4] The emerging strategy is based on a recognition that if Sweden fulfills its green convictions in a way that prices Swedish industries out of world markets, then production and jobs will simply move to environmentally irresponsible nations, negating both the environmental and economic goals.

Bowing to economic necessity, the 1991 energy legislation reversed past policy on nuclear decommissioning, declaring, "The start of the [nuclear] phaseout shall be determined by the success of efforts to increase energy efficiency, by the availability of new, environmentally acceptable means of power production, and by the goal of maintaining internationally competitive electricity rates" (Swedish Institute 1991).

Priorities of the new policy that have potentially important agricultural impacts are energy conservation and development of renewable, "environmentally acceptable" energy sources. The 1991 energy legislation earmarked more than US$600 million to be spent over five years on conservation measures, research, and development of renewable biomass, wind, and solar alternatives.[5] To facilitate renewable energy development, the food policy's land-conversion program permits grain production on contract to ethanol producers, and a portion of the conversion budget (equivalent to 100,000 hectares) could be used for construction of pilot ethanol distillation plants (MOE 1990b).

As part of the general tax reform, energy purchases were made subject to the general 25 percent value-added tax. In January 1991, additional green taxes were imposed on fuels that release carbon, sulfur, and nitrogen emissions. The carbon tax alone was expected to

[3] Nuclear power plants currently supply half of Swedish electricity.

[4] Energy costs are equal to 9 percent of value added in the paper products industry and 8 percent in mining and quarrying (Taylor 1992a).

[5] The budget breaks down to $167 million each for conservation and bioenergy plus $294 million for research and development of prototypes in wind and solar power.

increase yearly industrial costs by $1 billion and inflate the user costs of petroleum and coal by 44 percent and 128 percent, respectively.[6] A commission was assigned to forecast the combined effect of energy and emissions taxes on industrial production costs, exports, and employment. On the basis of a prediction that the taxes would devastate key export industries, including paper, mining, and metals, it recommended a 75 percent reduction in the general energy tax for industry. This recommendation was enacted in 1992, to the vexation of consumers and environmentalists.[7] The loss of state revenues from the energy tax cut is expected to be $650 million per year. Sweden currently imposes a 5 percent value-added tax on industrial energy use, versus 25 percent on consumer and public-sector purchases, plus the previously mentioned taxes on carbon, sulfur, and nitrogen. It is hoped and expected that other industrial nations will soon follow suit (Karlsson 1992, Taylor 1992a).

This high-stakes political contest is the setting for a discussion of agriculture's potential contribution to energy self-reliance and reduced greenhouse gas emissions. Biofuels, mostly from forest products and primarily used within the forest products industry, currently account for 15 percent of the total Swedish energy supply, or about the same as hydroelectricity. Bioenergy is exempt from the carbon tax, since carbon dioxide is recycled via growing plants. This economic edge has been augmented by the availability of payments for conversion of arable land to energy species (43,000 hectares) and by the grants for investment in pilot biomass energy plants already mentioned. The agriculture minister, Karl Erik Olsson, a dairy farmer and Central party member, has been assigned the biofuels portfolio. He is strongly committed to expansion of biofuels and advocates permanent price supports for energy crops.

The future extent and the mix of energy species is impossible to predict. Under the most aggressive promotional policy, they might occupy several hundred thousand hectares and contribute 2 to 3 percent of the domestic energy supply in the year 2015.[8] As the National Environmental Protection Board has observed, "It is economic prereq-

[6] The 1991 fuel taxes translate to $41 metric ton of CO_2, $5/kg of sulfur, and $6.70/kg of nitrogen.

[7] Swedish energy taxes add $23 to the cost of producing a ton of basic steel ($60/ton for high-value specialty steels). The comparable German and British taxes are $0.65 and $1.50/ton, respectively (Taylor 1992a).

[8] These figures are an interpretation of projections in NEPB 1990d. They represent agriculture's net energy contribution, after deducting the energy used to produce, handle, and transport bioenergy feedstocks.

uisites rather than physical resources that will impose limits on the use of biofuels over the next few decades" (1990d:44). Landscape considerations are another limiting factor, and conflicts have already arisen. It is intuitively obvious that the future economic competitiveness of agrobioenergy will depend on world price trends for petroleum and natural gas, on the level of green taxes on competing feedstocks, on technological developments in all types of energy generation and distribution, and on decisions regarding price support for energy crops.

Given present technology and price-cost relationships, burning grain straw in industrial boilers seems to be the most competitive form of agricultural bioenergy, even when the opportunity cost of removing potential soil organic matter is taken into account. Straw has the advantage of being a waste by-product of an existing farm production system, but it could never supply a nationally significant amount of energy. Burning grain is another possibility that requires no changes in farm production; indeed, wheat grown under contract to biomass power plants has received the acreage-conversion subsidy.

The available evidence indicates that ethanol is never likely to be profitable under Swedish climatic conditions. The cost of producing feedstocks (grain or sugar beet) is so high that they would have to be heavily subsidized. On the other hand, small-scale biogas facilities based on grass crops, possibly fermented with farm and household wastes, have attracted interest and research funding. Biogas appeals to environmentalists as a relatively clean fuel whose production also creates a high nutrient residue. With the present price-cost relationships and available technology, however, biogas generated from cultivated grasses and distributed locally still costs about twice as much as delivered natural gas. Silvander (1993) has shown that if "free" waste materials were available and a sufficiently high social value were placed on preserving the open landscape, small-scale biogas production could nonetheless be cost-competitive (Holstein and Åkesson 1991, Karlson 1992).

The greatest political interest in agricultural biofuels centers on energy forest, principally willow (salix) harvested on a four-year cycle from ratooning rootstock, and hybrid aspen grown on a short rotation of twenty to thirty years. In simulation studies, chips from woody species grown on fertile land with existing agronomic knowledge and equipment technology, and delivered to nearby biomass boilers, were nearly profitable at 1989 petroleum prices. When the new carbon tax is factored in, the unit cost of heat generated from salix is actually

Figure 9.1. Harvesting salix for bioenergy (photo by Bo Gustavsson, courtesy of *Landbild*)

lower than from oil or natural gas. Only coal and biomass from conventional forest operations have lower costs.

On the production end, the $250,000 capital cost of a *salix* harvester/chipper (Figure 9.1) requires operations of several hundred hectares to be economically feasible. But machinery cooperatives or contract harvesting, which are common in Swedish forestry, could take advantage of scale economies without requiring much larger farms. A second key economic factor is over-the-road hauling costs for wood chips. Unless more efficient bulk transport is developed, chip-burning power plants would have to be located near crop-producing areas. This favors biomass production close to population centers with substantial demand for thermal energy (Parikka 1989).

Since the economic viability of bioenergy species requires both high yields and nearby demand, bioenergy has little potential to sustain agriculture in sparsely populated mixed and forest regions. Most bioenergy crop plantings would be in medium- to high-yield grain-producing areas near central Sweden's major population centers. The most fertile southern regions would continue to specialize in cash grain and intensive livestock production. A high tax on carbon dioxide would indirectly benefit agriculture in disadvantaged regions, especially dairying, if bioenergy became profitable enough in the central Swedish plains to displace dairy farming there.

Forecasts of future energy crop acreage on arable land range from tens of thousands to one-half million hectares, or one-sixth of all arable land. Our best guess is that 100,000 to 200,000 hectares (3% to 6% of arable land) will be converted to some bioenergy species by the end of this century, if several conditions hold: world oil prices go no higher than US$30 to $35 per barrel, the current tax on carbon emissions is retained, energy crops do not receive price supports, biofuels research and development expenditure is at the planned level, and no further subsidies are made for bioenergy power plants. Perhaps the most critical unknown is what the European Community's carbon tax and biofuels policies will be in the latter years of 1990s.

The current research efforts to improve the efficiency of energy crop production, harvesting, transport, and power generation are likely to have a high long-run social return, so public funding is warranted. But on the basis of present knowledge, price supports for crops that cycle carbon and investment subsidies for biomass energy plants are not justified. Basically, appropriate green taxes on nonrenewable and polluting energy sources are a more cost-effective and more equitable incentive mechanism.

As a positive economic spinoff from agrobioenergy development, Sweden might well gain a competitive edge in a growing international market for bioenergy-cropping equipment and energy-generating technologies. This would build on earlier Swedish export success in such fields as industrial emission controls and nuclear waste disposal (Porter 1990:140).

The environmental benefits of producing grasses and woody species for bioenergy would be a net reduction in carbon emissions, less nitrogen leaching, and the addition of another way to recycle sewage sludge.[9] But policymakers must weigh three negative implications of having large tracts of salix and aspen on arable land: visual blight, diminished recreational access, and reduced land-use flexibility.[10]

[9] Note that our forecast of the likely location of energy crops suggests that little of the leaching reduction would be in areas where leaching is a serious problem.

[10] Salix is typically harvested on a four-year cycle, but its high establishment costs for field preparation, rootstock, and planting mean that it takes several harvests to amortize the initial investment.

Dismantling the Agricultural Bureaucracy and Privatizing Extension Services

Agriculture is a target in the bourgeois government's crusade to scale down, decentralize, and privatize state functions. In 1991, the National Board of Agriculture and the Agricultural Marketing Board were fused into the smaller, more streamlined Swedish Board of Agriculture. The county agricultural boards were replaced by agriculture departments within the general county administration and their staffing was cut. There have even been rumors that the Ministry of Agriculture will be subsumed within the Ministry of Industry, though this will not happen while the Center party is part of government.[11] In brief, the state has renounced its historic guardianship of the farm sector: the bureaucratic side of the iron triangle is buckling.

Agricultural extension services are no longer considered an appropriate state function. Technical, commercial, and financial advice is seen as a suitable sphere for private enterprise and producer cooperatives. According to Nitsch (1991), the new county agriculture offices will concentrate on planning, control, and implementation of political decisions, leaving few resources for advisory work. This path to privatization of extension was paved during the late 1980s when the county agricultural boards began to charge user fees for some commercial farm advisory services. Public grants to the local agricultural societies have also been pared down, so extension's noncommercial functions are dwindling as well.

Summary

Some of the measures described here were designed to complement the agricultural policy reform, but others were undertaken with little or no heed for their impact on the farm economy or the rural environment. Taken together, the most pronounced effects are likely to be:

- A further decline of farming in economically disadvantaged regions, due to higher tax burdens and more competitive commodity pricing.
- Fewer small-scale dairy farms, due to costly manure storage requirements in the southern counties.
- A further small reduction in nitrogen leaching and cadmium buildup in soils.
- A modest expansion of bioenergy production, stimulated by a mix of land-conversion subsidies, public R&D investments, and the tax on

[11] This action will probably be deterred by the Center party's opposition and by anticipation of joining the EC, where national agriculture ministries remain prominent.

carbon emissions. Bioenergy from short-rotation energy forest will reduce pollution but will have an adverse effect on the landscape.
- Lower retail food prices and a slight increase in food demand, due to more intense competition in food processing and distribution.

In sum, the impact of recent nonfarm policies on agriculture, rural regions, and the environment is mixed.

9.2 Swedish Agriculture in a New World Order

Up to this point, the effects of Sweden's new food policy and other recent legislation have been discussed as if no external changes would complicate the outcomes. That premise is clearly invalid, and here we touch briefly on the possible effects of four changes of great import: the end of the Cold War, a GATT treaty reducing agricultural trade distortions, the increasing green conflicts in international trade, and Swedish membership in the European Community.

Food Security in the Post–Cold War Environment

Since the end of World War II, the national security rationale for agricultural protectionism and for special Norrland supports has been founded on low-probability, high-risk scenarios centering on disruption of access to imported staple foods and farm inputs and, in some scenarios, disruption of internal distribution as well. We have seen that even before the collapse of the Soviet empire and dissolution of the Soviet Union, strategic thinking about Swedish food security was shifting away from the past emphasis on peacetime self-sufficiency and maintenance of production capacity in all regions. Stockpiling food and critical farm inputs took on greater importance in crisis planning.

Although the threat of a World War III has receded, the food security issue has not been permanently laid to rest. Political and military stability in nations just a stone's throw across the Baltic from Sweden is not ensured. As Defense Minister Anders Björk recently put it, "In a long-term perspective, there is obvious uncertainty, particularly with regard to developments in Russia and its neighboring countries" (cited in Taylor 1992b).

The 1991 Persian Gulf war was also a reminder that the heralded "new world order" is likely to be disorderly in ways that could jeopardize international flows of critical resources, including petroleum and conceivably even fertilizer and food. When such long-term ecological uncertainties as climate change, sea-level rise, and ozone thinning over high latitudes are taken into account, it seems to us that food security and renewable energy should still be on the policy agenda.

For years to come, there will be debates about optimum levels of food and bioenergy production, regional agricultural capacity, and strategic stockpiles. But the context of these debates and the constraints on policy choice are likely to be profoundly affected by a GATT resolution on agricultural trade and Swedish membership in the EC. Indeed, the end of the Cold War was a catalyst for fundamental rethinking of Swedish diplomatic-military neutrality, the biggest stumbling block to EC membership.

Outlines and Effects of a GATT Resolution

The Riksdag's food policy debate in the latter 1980s was premised on the assumption that a Uruguay Round treaty would soon restrict nations' freedom to impose trade-distorting agricultural policies. In general, Sweden has sided with the United States and the Cairns Group in advocating major reductions in price-based farm support, import restrictions, and export subsidies. In fact, just before the first breakdown of negotiations in 1990, Sweden's Minister of Agriculture Matts Hellstöm attempted to broker a compromise between the Cairns Group and their EC and Japanese opponents.

In 1992 the United Sates and the European Community resolved the "oilseed war," which seemed to be the final major stumbling block to a compromise formula. There were further promising developments at the June 1993 G-7 economic summit in Tokyo. Nonetheless, as of this writing, a GATT agricultural agreement has not been finalized. It appears that, if and when it is reached, the compromise solution will resemble the proposal offered in late 1991 by GATT's director-general, Arthur Dunkel. Variable import levies will be replaced by fixed tariffs, export subsidies will be restricted, income support will gradually be decoupled from farm prices, and overall budget support to agriculture will be cut one-third or more by the end of the 1990s (this is not an especially radical plan, since analysts predict subsidized exports would be reduced by 25% at most) (Viatte and Cahill 1991).

Sweden is well on its way to implementing the Dunkel Plan's conditions, and the government would be prepared to accept an even more sweeping deregulation of domestic and world food markets. Based on limited analysis of how such a gradual and partial deregulation would reduce allowable tariff levels and eventually increase international prices, we would expect most of Swedish agriculture—with its highly developed internal transport and distribution systems—to remain competitive in home markets through the end of this century.[12] Imports of some animal products, especially cheese, could increase significantly. Much would depend on whether the EC's and United States's stockpiles of surplus commodities could be disposed of without depressing world prices below Swedish production costs for several years and thus driving efficient farms out of business. A GATT treaty along the lines of the Dunkel Plan would accelerate the agricultural decline in Sweden's economically marginal regions, especially if it required the abandonment of Norrland price supports (see Blandford et al. 1990).

Toward a GATT Green Round?

As public concerns about environmental quality and food safety increase in all the industrial nations, agricultural trade controversies will inevitably persist in new forms after completion of the Uruguay Round. Ad hoc import restrictions justified on grounds of food safety, animal rights, and environmental protection have become a hotly contested international policy issue. Indeed, a "green round" to resolve these issues has already been proposed (Uimonen 1992). Carol Kramer identifies the economic stakes that lie behind a series of past green trade disputes, ranging from the EC's import restrictions on hormone-fed U.S. beef to the United States's restrictions on alar-sprayed Chilean apples: "One of the major agricultural proposals being negotiated under the GATT is the harmonization of national food safety standards. The motivation for globally establishing and maintaining the same food safety standards is that divergent national standards have been used as nontariff barriers to trade. *There is also a fear that countries will increasingly resort to these standards if and when other trade barriers are eliminated*" (1991:10; emphasis added).

In the area of product safety, there appears to be universal agreement that the Codex Alimentarius and the GATT's Article XXb as they now

[12] Blandford et al. (1990) project that complete elimination of import protection and export dumping (under 1990 supply and demand conditions) would raise the international price of milk by 27 to 61 percent and coarse grains by 10 to 30 percent.

stand are inadequate for establishing international standards as well as resolving international disputes. Reaching an accepted set of standards will not be easy, however, because governments, economic interest groups, and environmentalists disagree fundamentally about what are legitimate purposes and what role scientific criteria should play in standard setting and dispute resolution.

Difficult as it is to reach agreement on product-safety criteria, this is only one aspect of a much larger challenge: to establish international standards for environmental protection and animal rights, and to specify the legitimate powers of nations with especially stringent standards to protect or compensate their producers for high compliance costs. A resolution of these technically and economically complex—and politically divisive—issues will not come soon and will undoubtedly require at least one major round of international negotiations in the 1990s (Runge and Nolan 1990).

Swedish agriculture could be profoundly affected by the outcome of such negotiations. The economic competitiveness of Swedish crop producers would be enhanced by other countries' adoption of measures that either raise the price or reduce allowable quantities of pollution-causing farm inputs. Since Sweden's main crops—small grains and fodder crops—are not ones involving intensive fertilizer and biocide use in other countries, however, this cost advantage would be fairly small (Tobey 1991). Indeed, Swedish competitiveness would be adversely affected in a free trade regime that did not allow some form of protection or compensation for the added costs linked to Sweden's tough manure handling requirements, green input taxes, cadmium limits, and animal-welfare requirements.

Swedish livestock producers would benefit if the rest of the world adopted its high standards for manure management and animals' well-being. Inversely, the most serious threat to Swedish agriculture would be a free trade regime with no allowance for protection or compensation based on humanitarian and environmental objectives in the livestock sector (Jägerhorn 1992a).

Sweden has a stake in creating a common negotiating front with such nations as Denmark, Japan, and Switzerland, which also have high green standards for agriculture. This would almost certainly require a willingness to modify the instruments and standards Sweden has introduced unilaterally over the past several years. In our view, such a sacrifice of national sovereignty would be in Sweden's interest, since the payoff—more rigorous green standards in other advanced industrial nations—would both improve Swedish economic competitiveness and fulfill its green commitments.

Sweden in the European Community

Within a year of Sweden's food policy reform, the government submitted its application for full EC membership. This would be Sweden's most profound political and economic policy redirection of the twentieth century. It is reflective of the country's political change and instability that two actions with such momentous and potentially divergent consequences for agriculture—the food policy reform and the EC application—were taken in the same year.

One might ask, as Ewa Rabinowicz (1991b) has, whether it was rational to initiate large-scale and in some ways irreversible changes in Sweden's farm economy only to subject it to a new type of regulatory regime within a few years. Notwithstanding such doubts, the newly created Swedish Board of Agriculture has adopted the following motto: "We shall work for competitive, environmentally friendly, and livestock-friendly food production in a European perspective."

As background to formal negotiations, which began in January 1993, the EC Commission communicated the principal policy changes it considers necessary for Swedish membership. To no one's surprise, the commission stressed agricultural and regional development policies.[13] Negotiations should be completed by the end of 1993, with membership possible by the end of 1994, depending on the outcome of a Swedish popular referendum. The timetable may be set back by the present currency instability and by delays in adoption of the Maastricht Treaty for European Union. It would certainly be affected if the defeat of Maastricht precipitated a political crisis among the twelve EC members. Even its defeat, however, would not likely spell the end of the Common Agricultural Policy or dissuade the Swedish government from seeking membership.

Perhaps a more serious obstacle is Swedish public sentiment, which is deeply divided about EC membership, even though all four parties in the coalition government, plus SAP's leaders, support it. The negative result of the Danish referendum on Maastricht in June 1992 at least temporarily dampened popular enthusiasm about EC membership in all the Nordic countries. In October 1992, 53 percent of Swed-

[13] Also emphasized were differences between Sweden and the EC in social and alcoholic beverage policy and, of course, Swedish military/diplomatic neutrality (DN 1992).

ish voters opposed and only 30 percent supported EC membership. After the Danes approved a revised version of Maastricht in early 1993, Swedish voter sentiment began to shift. In March 44 percent opposed membership, with 34 percent in favor and 22 percent still undecided (Berg 1993).

The central question in attempting to forecast the effects of EC membership on Swedish agriculture and environment is what the Common Agricultural Policy itself will look like in the future. The CAP has been under ceaseless pressure because of its spiraling budgetary cost, its manifest economic irrationalities, and its status as the most serious obstacle to a Uruguay Round agreement. In 1991, EC Agricultural Commissioner Ray MacSharry proposed a set of measures to remedy these deficiencies. The MacSharry Plan would to some extent decouple farm income supports from commodity prices and production levels. It would replace variable import levies with fixed tariffs and compensate farmers for income and capital losses by making payments based on acreage and numbers of livestock. These actions, would permit a reduction in farmgate prices, import levies, commodity storage costs, and export subsidies.

In May 1992, after seventeen months of intense lobbying and debate—and over farmers' strenuous protests—the EC's agricultural ministers agreed to the most far-reaching changes in the CAP's history. (With only Italy dissenting, there were sufficient votes to implement the reform under the EC's new formula for weighted majority rule.) The new policy calls for small grain and beef prices to be cut by 29 percent and 15 percent, respectively, over three years, starting in 1993. Farms larger than twenty hectares must also set aside 15 percent of their cropland to reduce excess production still further. The volume of surplus meat purchased and stored at EC expense will be halved, from 750,000 metric tons in 1993 to 350,000 in 1997. Milk quotas and the intervention price for butter will also be reduced slightly. In return, producers of grains, oilseeds, and protein feeds will receive acreage-based compensation, and beef producers will receive compensation per animal (Höök 1992b). The policy is, of course, far more detailed than we can describe here. The price reductions, less drastic than those urged by MacSharry, represent a compromise that will reduce but probably not eliminate commodity surpluses. The compensation-plus-set-aside mechanism sufficiently resembles the United States's deficiency payment approach to suggest that one of its main purposes was to break the Uruguay Round deadlock.

In speculating about the economic and environmental impacts of

Swedish EC membership with a CAP reformed along lines of the May 1992 decision, four factors are central:

- The levels of CAP import barriers, export subsidies, farmgate prices, and income support
- The impact of exposure to community-wide competition in agriculture and especially in food processing and distribution
- The EC's regional development policy and revenue-sharing plans for economically disadvantaged regions
- The EC's movement toward common environmental protection standards and measures, plus its revenue sharing for environmentally sensitive areas.

If Swedish farmers continue their adjustment to lower farm prices over the coming two years, major segments of a "lean" grain sector in the southern plains and dairy and horticultural sectors near major population centers should be able to withstand external competition for the domestic market. If there are parallel price reductions in Sweden and the EC in 1993 and 1994, small grain prices should be roughly equalized by the time Sweden enters the EC in 1995. For fluid milk, handling and transport advantages should insulate efficient Swedish producers near large population centers, except possibly those just across the Öresund from Denmark. Beyond the CAP's pricing policy, the factors that will most affect Swedish agriculture are the levels of income support, acreage and livestock payments, and the quota of total EC milk production allocated to Sweden. One competitive advantage for Swedish farmers is that by 1994 they will have adjusted to life without income supports.

Small and part-time farms that rely primarily on local direct market outlets or that produce specialty and value-added products should also be able to maintain their markets and profitability. This raises the question of consumer loyalty based on nationalistic sentiments, solidarity with local farmers, perceptions of superior product quality, or a desire to sustain agriculture's nonfood amenities. The LRF has already intensified its promotional campaign to persuade Swedish consumers to stay with "clean and green" local produce (implicitly: at a premium price). Evidence from the studies cited earlier suggests that such appeals will have some effect, but it is difficult to quantify.

Sweden's environmentally friendly reputation could also pay off in export opportunities. The LRF is convinced that with aggressive promotional and marketing efforts, Sweden can even carve out a sizable niche on the Continent for reduced-chemical and organically grown foods. Livestock producers may also find a profitable European niche for meat from Sweden's "happy pigs."

In contrast to these optimistic prospects, the competitiveness of Swedish oilseeds, cheese, and most meat products is likely to be severely eroded by low-cost imports. The expected decline in large-scale poultry and hog operations could be mitigated slightly by cheap feed grain and protein-feed imports and by the costly manure management requirements recently imposed by Denmark and the Netherlands, the toughest competitors for Swedish markets (described in Chapter 10).

The fate of Norrland agriculture, once deprived of high milk prices, looks especially gloomy. Production costs in Sweden's Norrland and the subarctic regions of Norway and Finland are far higher than those in most of the EC's present disadvantaged regions. The Norrland's agricultural future hinges on whether prospective new EC members win the right to extend special support to these far northern regions (Lerner 1992, NBA 1991c).

Sweden's high-cost food processors will be harder hit by EC competition than will farming per se, but as noncompetitive processors go under, many dependent farmers will go with them. A recent report commissioned by the LRF indicates that EC competition would require Swedish processors to cut their unit costs by 25 to 30 percent on average to maintain their current domestic market share.[14] To improve competitiveness, the report recommends merger and consolidation of small processors and distributors, greater capacity utilization, and joint ventures with counterparts in other EC nations. To this list we would add a need for more imaginative and aggressive marketing. Accomplishing these requisites for survival implies a more entrepreneurial management style than the farmers' cooperatives have historically provided (LRF 1992).

Despite this generally pessimistic prognosis, LRF's leaders support Swedish EC membership. Since they expect worldwide agricultural trade liberalization in any event, they believe Swedish farmers' economic interests can be better guarded within the CAP, where agricultural interest groups still have substantial political clout, than in an isolated, free-trading Sweden. LRF's members, however, are deeply split over EC membership.

[14] Most processors are technologically up-to-date but nonetheless economically inefficient. Their comparatively high costs are due primarily to small-scale, ineffective organizational structure, low capacity utilization, and high labor costs. Under the old policy regime, import protection and monopolistic markets permitted this pattern to persist, at the expense of consumers.

David Harvey (1991) predicts that as the EC moves toward internal agricultural policy reform and international trade liberalization, the priority given to rural socioeconomic cohesion and environmental protection will rise. Such a tendency would obviously parallel the evolution of Sweden's goal structure. As an EC member, Sweden would strengthen the community's environmentally progressive forces. Indeed, Denmark's foreign minister publicly welcomed Sweden's application for exactly this reason.

Austrian, Finnish, Norwegian, and Swiss membership would further strengthen the green alliance in EC politics. If all these economically advanced nations join the community, we would expect its agroenvironmental standards to converge toward Sweden's. The May 1992 CAP reform package already points in this direction, with its measures to support chemical-free agriculture, establishment of permanent pastures, and improvement of landscape quality (*Land* 1992).

There is a probable downside, however. The EC offers large-scale financial support for afforestation, including conifer planting. This would accentuate a shift in land use that most Swedes strongly oppose. And in the case of several specific measures and standards, such as green input taxes and maximum cadmium levels, Sweden will either have to accept different instruments and relax its standards or face higher production costs than its neighbors. The EC does not generally prevent member nations from imposing tough environmental regulations as long as they do not disturb intra-EC trade. But a nation's attempts to compensate farmers for their high compliance costs are certain to be protested as hidden subsidies.

Green harmonization was expected to be challenged by Sweden's strict animal husbandry standards. Because the purpose of the standards is to protect farm animals' well-being, Sweden would presumably seek to restrict importation of animal products produced under inhumane conditions. In practice, Swedish negotiators have taken the pragmatic path in not contesting this issue. Rather, the State Board of Agriculture's strategy, backed by the LRF, is to maintain high husbandry standards in domestic livestock production and publicize them to gain a competitive advantage with Swedish and foreign consumers.

Conclusion: Internationalization and the Future of Swedish Agriculture

Sweden's market-oriented food policy will accelerate structural and regional polarization. The economic dominance of large-scale farms in the flatlands of south and central Sweden will be accentuated, as will the demise of smaller commercial family farms and all types of farming in the mixed and forest regions. In those areas, coniferous forest will most likely invade much of the remaining open landscape within a relatively short time. The tax and antitrust laws enacted since the food policy reform have a similar thrust. Of course, these long-term trends existed under the old farm policy as well, but they were mediated by its special provisions. In the absence of a GATT treaty and Swedish EC membership, these trends would continue to be blunted somewhat by Norrland price supports, rental payments for socially and biologically valuable landscapes, and subsidies for deciduous tree planting.

Our expectation is that a GATT treaty and Swedish participation in a reformed Common Agricultural Policy would strongly reinforce these structural and regional trends, not least by eliminating the milk-pricing policy that has kept Norrland farming alive. In such a policy environment, it is problematic whether the CAP's milk quotas, acreage payments, and structural measures for environmentally sensitive and economically disadvantaged areas would be a sufficient counterforce to "survival of the fittest" competition. In a free trade environment, it is also problematic, for both legal and economic reasons, whether Sweden could cling to its high standards of environmental protection and animal welfare. Some unpleasant trade-offs may be required.

Opening up the food system to European and global competition will enhance Swedish consumers' material well-being and improve narrow economic efficiency in food production and distribution. These have been the political parties' prime objectives since the latter 1980s. But Swedes will be worse off as consumers of the amenities associated with open and varied agrarian landscapes and with rural communities that retain elements of a fondly remembered agrarian way of life. Any hope for a greener agrarian future than this hinges on Sweden's ability to forge alliances with other environmentally progressive nations—in the EC, the GATT, the UN's Sustainable Development Program, and other forums—to move worldwide agriculture in a greener direction.

10

Universal Greening: International Patterns

Negative agroenvironmental symptoms, ideological greening, and demands for a more environmentally friendly agricultural policy are now universal in the advanced industrial nations. We believe that Sweden's agroenvironmental initiatives have been more rapid and more comprehensive than those of other nations. Nonetheless, broad patterns of convergence are becoming apparent; furthermore, other nations have been the innovators in many aspects of greening.

Universal greening is reflected in the explosion of related literature over the past few years. There have been several multicountry studies comparing policy responses to a range of environmental problems (Dubgaard and Nielsen 1989; EC Commission 1990; Ervin and Tobey 1990; Hanley 1991; Haxsen 1990; OECD 1989b; Young 1988, 1989). Other comparative studies concentrate on single issues such as land use, wildlife habitat, or nitrogen pollution (Baldock 1990, Cloke 1988b, Hanley 1990, Hurst 1991, Whitby and Ollerenshaw 1988). And countless more works analyze individual nations' problems and responses.

An issue-by-issue or country-by-country recapitulation of all these publications and policy initiatives is not possible. Instead, this chapter describes five broadly similar greening tendencies, highlighting national initiatives that seem particularly innovative, well conceived, or internationally influential. It also considers why Sweden, whose agroenvironmental problems are less severe than those of many other countries, has been in the vanguard of green experimentation, and it returns briefly to the forces that may promote or obstruct international harmonization of agroenvironmental policies in the 1990s.

We address green agricultural policy initiatives only in advanced capitalist nations, with no systematic East-West or North-South comparisons.[1] That does not mean that agriculture's detrimental effects on the environment are less acute outside industrial capitalism. On the contrary, an outpouring of recent revelations confirms that soil and water degradation, not to mention farm-worker safety and diet-related health problems, are severe in the transitional economies of eastern Europe and the former Soviet Union. It is problematic whether the serious agroenvironmental contradictions inherited from the communist period will be treated with urgency in the immediate future, because most postcommunist regimes have given priority to economic restructuring and productivity growth in their farm sectors (Brown and Young 1990, Csaki 1990, French 1990).[2]

North-South contrasts are, if anything, even more stark. Scores of Third World nations remain entangled in a web of rapid population growth, urban-biased development, widespread chronic malnutrition, dependence on cash crop exports, and foreign debt. There is persuasive evidence that the long-term carrying capacity of agroecosystems is being undermined in large parts of Asia, Africa, and Latin America. The urgency of beginning the transition to an ecologically sustainable agriculture is, if anything, far greater there than in the affluent North. The 1992 Earth Summit in Rio de Janeiro helped educate world citizens to the fact that North-South relationships have been part of the problem and must certainly be part of the solution. Sahelian desertification has been exacerbated by misguided foreign-aid projects for livestock and groundwater development. Tropical deforestation is accelerated by hell-bent agricultural export development, made imperative by heavy debt-service obligations. Rural poverty in nations that depend on exporting staple food commodities is made more intractable by rich countries' export dumping. And groundwater poisoning in parts of Asia stems from the uncritical adoption of a Green Revolution technology grounded in Western science and promoted by

[1] We had hoped to bring material in this chapter up-to-date by making use of an OECD volume on agricultural and the environment, scheduled for publication in 1992. Because of that work's publication delays, we have relied heavily on the earlier OECD survey (1989b), now largely overtaken by events. The rapid pace of policy changes and the explosion of knowledge generally will quickly render obsolete much of the detail in the following narrative.

[2] Probable exceptions to this forecast are Hungary, Bulgaria, and Romania, all net exporters of farm products and processed foods. They will have to be convincingly "clean" if they want to maintain or expand food exports to Western markets.

agribusiness input producers (Brown and Young 1990, Ruttan 1990, Southgate et al. 1990, World Bank 1986).

10.1 Differences within Similarity

A 1989 OECD survey, *Agriculture and Environmental Policies,* concludes:

> There is a remarkable consistency in the kinds of environmental problems associated with agriculture that have been inventoried in many OECD countries, although their wide geographic and climatic diversity means that the relative importance of different environmental problems . . . varies widely. Animal manure in semi-arid *Australia* is not the problem that it is in sub-arctic *Finland.* Problems associated with the density and proximity of urban and agricultural lands in *Japan* and the *United Kingdom* are far less critical in sparsely populated *Canada* and *New Zealand.* Problems of agrochemicals are more acute in European countries with high population densities [and] intensive agriculture than they are in countries like *Canada,* the *United States,* and *Australia* where . . . more extensive practices [are] economical. (1989b:82)

In the latter half of the 1980s there was a groundswell of popular demands for a more environmentally friendly agriculture. In their own distinctive ways, many societies experienced catalytic forces similar to Sweden's:

- Growing affluence accompanied by increased demand for environmental amenities and food and water free of harmful chemical residues
- Intensified media attention to farm policy's negative impact on the agrarian landscape, rural communities, water quality, and animal well-being
- The growing involvement of broad-based environmental organizations in agricultural issues and their development of ties to the alternative agriculture and historic preservation movements
- Continuing encroachment of suburban, recreational, and tourist development into traditional farming regions
- Widening regional economic disparities, caused in part by the productivity gap between advantaged and disadvantaged farming regions and by the disappearance of support infrastructures in marginal areas
- Increased scientific understanding and popular awareness of industrial agriculture's local environmental degradation and the threat to long-

run sustainability posed by climate change, ozone depletion, and acid deposition
- Accumulated frustration about the power of iron triangles to obstruct meaningful agricultural policy reform.

In addition to the presence of similar causes, there were signs of convergent responses. Ingemar Nilsson observes: "Sweden has in most respects stricter rules than EC nations, as applied to both animal protection and crop production's environmental impacts. At the same time, it is obvious that the EC nations, at least those in northwest Europe, are going through the same development, even if on average they are somewhat later in phase" (1989:4).

We have interpreted Swedish greening in the 1980s as an accumulation of incremental policy revisions, not a radical reform. By implication, initiatives in other countries also fall short of radical reform. Nonetheless, in the span of a few years, many nations have taken measures comparable to Sweden's: manure storage and spreading requirements, tighter biocide registration procedures, integrated pest-management programs, more research on reduced chemical techniques, cross-compliance stipulations in acreage set-asides, multiyear landscape-management contracts, afforestation subsidies, organic certification standards, agricultural bioenergy experiments, animal-rights legislation, and so on (see Dubgaard and Nielsen 1989; Ervin and Tobey 1990; Hanley 1991; Haxsen 1990; OECD 1989b, Young 1988, 1989).

To some extent, convergent policies result from deliberate attempts at harmonization. The EC's Common Agricultural Policy, in particular, has sought greater consistency among member nations through directives on the use of livestock medications and growth stimulants, land-management agreements in "environmentally senstive areas," "extensification" (reduced input intensity) tied to set-aside payments, and reduction of nitrogen leaching in "vulnerable zones."[3] The nations seeking EC membership face pressure to accommodate their policies to its green standards.

The creation of the EC Department of Environmental Protection in 1990 reinforces the commitment to a consistent community-wide pol-

[3] Ervin and Tobey (1990:15) view the EC's land set-aside program as first and foremost a supply-control tool, one that makes little tangible contribution to green objectives—or to output reduction, for that matter. Germany stands out for the large proportion of set-aside land that is subject to extensification measures. Through 1989, however, these methods were not used widely in the EC as a whole. As noted earlier, many analysts believe that fewer inputs on a fraction of farmland tend to be offset by more intensive use on remaining land.

icy, although Ervin and Tobey foresee difficulties grounded in divergent agricultural problems and priorities: "In northern Europe there seems to be a particular concern about ecosystem pollution and wildlife habitat. Southern European countries tend to stress the need for maintaining small marginal farms to prevent land abandonment and the deterioration of rural social infrastructure. Such divergent environmental issues necessitate an EC program structure that can be flexibly applied" (1990:25). Some northern European members, by this interpretation, "tend to regard the EC as a threat to their own [tougher] standards" (1990:13). In less-developed southern Europe, agriculture continues to be viewed in essentially economistic terms: as a major employer, a growth sector, and a source of foreign exchange (Bonnano 1989).

The nations of northern Europe have signed treaties in Paris and Helsinki to curb nitrate and phosphate pollution of the North and Baltic seas. And the Nordic countries, with their cabinet-level environmental council, consult closely on environmental legislation, share research findings, and have several specific measures in common. On a still broader front, OECD members have "reached a consensus that in future they will improve the integration of their agricultural and environmental policies" and move toward consistent application of the Polluter Pays Principle, which OECD adopted back in 1972 (1989:3).

The UN Food and Agricultural Organization's Codex Alimentarius Commission is yet another international arena, in which governments, food industries, and environmentalists contest health and safety standards, as affected by biological and chemical preparations in agriculture and food processing. The Codex's ostensible goal is international harmonization based on objective scientific criteria. But economic protectionism was close to the surface when a trade war nearly broke out between the EC and the United States in 1986–87 over the community's prohibition of imports of U.S. beef produced with synthetic hormones; similar motives lay behind the United States's boycott of Chilean grapes alleged to contain arsenic residues. These are early warning signals of a form of international conflict that seems certain to intensify in the 1990s (Kramer 1991).

The pattern whereby nations selectively adapt one another's policy experiments to their own conditions, with fairly short time lags, reflects the phenomenal expansion in international communications noted at the beginning of this chapter. Agricultural economists and life scientists, environmentalists and organic farming advocates, farm program administrators and legislators all exchange ideas and experiences with their peers through dense and expanding networks. Indeed, we view this book as a small contribution to the international pooling of knowledge. When information crosses borders so freely and quickly and when national policies have so much reciprocal influence, it is often difficult to distinguish the innovators from their imitators.

Our review of the literature suggests that five patterns are common to most industrial capitalist countries, including Sweden. First, agriculture's privileged position still persists when it comes to internalizing negative externalities: polluters seldom pay fully. Nonetheless, the social contract between farmers and society seems to be eroding. Second, agricultural and general environmental policies are becoming integrated, although political turf battles continue. Third, green policy measures have frequently been hasty and piecemeal, introduced without thorough review of the alternatives, probable outcomes, or inconsistencies with other policies. Fourth, many green initiatives have been tied to and justified as measures to curb excess farm production. Fifth, a greening of agroscientific research is under way with its prime objective being to find technological solutions for economic-environmental problems.

Greening and the Erosion of Agriculture's Privileged Status

Although agriculture's adverse environmental effects are gaining greater political prominence, its history of privileged policy treatment in fact stems from a "long *positive* association with the environment" (OECD 1989b:13; emphasis added). Katherine Reichelderfer, assessing the U.S. situation, concluded:

> Despite the potential of modern agriculture for harming the environment, agricultural activities in developed economies have been regulated quite differently with respect to environmental consequences than have the activities of mining, manufacturing, energy, and construction sectors from which similar environmental damages arise. . . . Unlike other sectors of the economy—in which pollution has increasingly been controlled through federal standards, fees and fines, restrictions and (more recently) market-based incentives—agriculture is unique in having engendered less federal government intervention with respect to its environmental consequences. When intervention has occurred, it has been achieved—more often than in other industries—through mechanisms that increase rather than decrease farmers' incomes. . . . While the centralized command-and-control approach has been given precedence in nonagricultural sec-

> tors, incentive-based and subsidy approaches have predominated in the agricultural sector. (1991:4)

A survey of agroenvironmental measures in Europe comes to a similar conclusion. Although lip service is paid to the Polluter Pays Principle, there is nonetheless a "strong preference for voluntary, subsidy approaches. When regulatory and enforcement measures are employed, partial or full compensation is usually provided. The cases of input charges (taxes) are few and then the revenues are returned to agriculture" (Ervin and Tobey 1990:i). Thus, Danish hog farmers may apply for government grants to cover 30 percent of the capital cost of meeting tougher manure storage requirements. Maize-soybean farmers in parts of the U.S. Midwest, where intensive rowcropping erodes up to fifty tons of topsoil per hectare yearly, are not required to curb their soil loss but may voluntarily enroll their land in the Conservation Reserve Program and receive compensation. Legal systems preponderantly give private landowners, rather than society, property rights over soil and water resources.

It appears that even when the command-and-control approach is used, regarding, for example, pesticide spray drift in France, earthing over manure in Germany, and planting fall cover crops in Denmark, the implementation of such measures relies heavily on advisory services and voluntary compliance. This voluntaristic bias reflects a fundamental reality: enforcing detailed regulations over a large number of geographically dispersed farm operations is costly and logistically difficult. The technical obstacle is especially clear for such nonpoint source pollution as nitrogen leaching and soil erosion (Dubgaard 1990b, Hurst 1991, OECD 1989b, van den Weghe 1990).

Another interpretation of the preference for voluntarism, consistent with Swedish experience, is that it is a remnant of the social contract between farmers and the rest of society. Many analysts confirm that politicians were reluctant to challenge property rights granted to "tillers of the soil" when agricultural pollution and resource depletion were less serious (or at least less well understood). Despite the persistence of goodwill toward farmers, there is a pervasive move to reconsider environmental property rights and bring agriculture within the scope of broad environmental protection legislation (Batie 1990, Ervin and Tobey 1990, Hodge 1991, OECD 1989b, Young 1988). Harvey believes this move is evidence that the social contract is eroding:

> This contract has been implicitly conditional on (i) the continuing acceptability of the socio-economic costs of agricultural support (tax costs and

> surpluses) and (ii) the continued care of the countryside as a largely agricultural countryside, at least in terms of land use if not socio-economic activity. The increasing erosion of the special position of agriculture then arises because (i) the socio-economic costs of traditional support are increasingly unacceptable . . . (ii) the care of the countryside has not been fulfilled—people have been leaving, land has been abandoned, countryside has been destroyed [and] (iii) the aims of the contract have shifted from securing food supplies to conserving rural social and natural environments.[4]

Incorporation into the Environmental Policy Framework

The OECD's 1989 survey on agriculture and the environment notes that a common tendency in the current third generation of environmental policy since the 1960s "has been toward the progressive transfer of authority from agricultural to environmental authorities" (1989b:26). Even when agricultural agencies retain operational authority for green measures, "a noticeable recent development has been the growth of mandatory inter-agency consultation procedures and formal inter-agency agreements for the development and implementation of policy" (1989b:26). From Austria to Australia, "committees, task forces, working groups and issue-specific inter-agency groups" are being created to oversee programs that run the gamut from protecting the environmentally sensitive areas to reducing surface water eutrophication (1989b:27). Great Britain's Countryside Commission, founded in 1968, is a venerable example of the quasi-autonomous, function-specific agency designed to operate at arm's length from the agricultural establishment. Canada's Soil Conservation Council is a more recent example (see 1989b:25–27, 84).

According to the OECD, "Austria, the Netherlands, United States, United Kingdom and Switzerland provide examples of a highly evolved institutional response to problems at the agriculture-environment interface. . . . The Dutch working group of three ministries which produces a revolving multi-year indicative program for agriculture and the environment" is singled out for special mention. Denmark's broadly representative Committee for Farming Policy is called "the most explicitly integrated mechanism" for coordinating the roles of public, quasi-public, and private actors (1989b:88).

It is clear that in many countries, restricting agricultural ministries'

[4] Personal communication from Dr. David R. Harvey, University of Newcastle on Tyne (1992).

jurisdiction over agroenvironmental concerns is a deliberate political stratagem. In the Netherlands, for example, pesticide and manure pollutants were causing locally severe degradation of water quality by the mid-1980s. The problem, and the policy failure that lay behind it, were evident and citizens were angry. Nonetheless, for several years the Dutch economic-bureaucratic iron triangle delayed or watered down measures requiring farmers to internalize the negative externalities. In a crisis management atmosphere, the state finally created special multiagency committees with strong mandates to attack nutrient leaching, denitrification, and pesticide pollution. There is ample evidence of similar foot-dragging by iron triangles in other nations (Hurst 1991, Tamminga and Wijnands 1991).

Growing intervention by environmental ministries also reflects the influence of environmental organizations and (in several countries) Green parties, as they have tackled agroenvironmental concerns and developed the necessary competence (Harvey and Whitby 1988).

Another reason for the shift in authority, in Sandra Batie's words, is that, "as a means of achieving environmental goals, agricultural policy is a blunt instrument at best" (1990:570). This helps explain why pesticide registration and water-quality monitoring have typically become functions of environmental protection agencies and why farmland preservation is often subsumed within general land-use planning at the national level (as in Japan, the Netherlands, and Switzerland) or regional level (as in Austria, Australia, and Canada).

In sum, the locus of both policy deliberations and formal authority is shifting. And yet M. D. Young concludes his survey of green initiatives in OECD nations by stressing that the "dominance of agricultural policy *instruments* over environmental instruments [continues] and, in particular, [there remains] a failure to develop integrated agricultural and environmental policies . . . [based on] conscious trade-offs . . . between competing objectives" (1989:4; emphasis added). The contest for jurisdiction is evidently not yet over.

Haste, Piecemealism, and Limited Information

It would be naive to expect green measures to emerge from the political process fully conceived, fully consistent with other policies, fully within the state administration's competence, and fully accepted by the affected parties. Rather, innovative measures will likely suffer from design faults and implementational obstacles ("bugs" and "glitches") and both instruments and implementation procedures will evolve through trial and error. These inescapable shortcomings are frequently exacerbated by the fact that green initiatives have been hastily concocted reactions to environmental bad news or demands for immediate action from environmentalists, the media, voters, and/or other nations.

There are several characteristic reasons for poorly conceived and poorly implemented green measures. First, the agroindustrial lobby uses its political leverage to dilute regulatory actions and to subvert the Polluter Pays Principle by securing compensation for environmental protection measures.

Second, the technical knowledge and empirical information needed for effective policy design are frequently underdeveloped, because of the irrelevance of much past research and the rush to take immediate action. Thus, when the European Commission moved to introduce a nitrate-pollution policy for the EC's "vulnerable areas," it discovered that the limited collection of relevant data in the past hindered timely and effective policy (EC Commission 1990). Similarly, the 1989 OECD survey notes that the bias toward ineffective voluntary measures and inefficient regulatory measures in pollution control is reinforced by limited information about economic instruments, such as regionally differentiated input taxes and marketable input quotas. Many countries have accelerated their efforts to identify more cost-effective and enforceable instruments (1989b:39).

At the scientific level, causal analysis is seriously underdeveloped. There is a "paucity of knowledge about linkages between certain environmental qualit[ies] and agricultural practices," ranging from how insects develop resistance to how routine antiobiotic dosing affects the health of laying hens (Batie 1990:570). Even more fundamental, however, is a problem found by Rod MacRae and his colleagues in Canada: "The development of political strategies in Canada to support the transition from conventional to sustainable agriculture has been limited by the absence of a comprehensive conceptual framework for identifying the most critical policies, programs and regulations" (1990:76).

Third, new divisions of authority and new working relationships among environmental, agricultural, and other agencies are seldom mapped out in advance. This impedes effective policy implementation and, as mentioned, bureaucrats and farmers in some countries have the will and the ability to resist or distort measures that challenge their autonomy or impose costs on farmers (Harvey and Whitby 1988).

A related problem arises when different political jurisdictions take conflicting approaches. John Holusha uses the expression "crazy quilt" to describe the mismatched standards applied by the fifty U.S. states in land-use regulation, property taxation, pesticide registration, and organic produce certification (1991). Reichelderfer further notes that "most major [U.S.] *farm* programs are Federal, while many *environmental* programs affecting agriculture are legislated and/or implemented at state and local levels" (1990:203; emphasis added). Indeed, in the Reagan and Bush years, the federal government lagged behind many states in developing alternative and sustainable agriculture programs.

Fourth, economic cost-effectiveness has not typically been high on the list of criteria used to select green instruments. We have noted the bias in favor of voluntary and command-and-control measures and the minimal use of economic levers that could lead to more efficient goal attainment. Politicians and bureaucrats seem slowly to be overcoming their ignorance of and aversion to market-like solutions.

As we elaborate below, policy fragmentation and weak instrument design seem to be especially common when green measures are patched into supply-control schemes. It is probably for the best that such measures are impermanent and that their funding depends on the severity of overproduction and budget constraints (see Ervin and Tobey 1990, Harvey and Whitby 1988, Hodge 1991, IAA 1991a, Weinschenck and Dabbert 1989, Young 1989).

Our evaluation of piecemeal greening in Sweden stressed the importance of learning by doing: fine-tuning instruments, streamlining implementation, and reducing policy inconsistencies over time. Such learning is widespread, and it is worth citing a few green initiatives that had evolved into coherent programs by the 1990s.[5]

Denmark and the Netherlands have put together comprehensive packages to mitigate their severe nitrogen and phosphorus pollution problems. Several of their early initiatives were in effect pilot projects for later Swedish measures (Dubgaard 1990a, b; Tamminga and Wijnands 1991; van Leeuwen and Oosternveld 1989).

Between 1985 and 1990, the United States coupled a broad range of resource-conservation inducements and requirements to its programs for acreage reduction and deficiency payments (see below).

Well before the mid-1980s, Switzerland and Austria devised a mix of incentives and regulations to maintain viable small, part-time farms and fragile, biologically rich pastoral landscapes in their tourist-depen-

[5] We cannot assess the overall goal attainment or cost-effectiveness of these measures.

dent alpine regions (Pevetz 1990, Young 1988). In addition, Britain, Germany, the Netherlands, South Australia, and other nations or regional jurisdictions have rural land-use plans and special measures to protect biotopes and farm landscapes (Baldock 1990, Cloke 1988a, Grossman 1989, Hanley 1991, van den Berg 1988, van Lier 1988).

Agroenvironmental policies everywhere are becoming more precisely calibrated to local variations in ecology, resources, rural culture, and farm structure. Norway's regional agricultural programs may be the most finely tuned of all, though reputedly at the cost of "a bureaucrat for every farmer" (Almaas 1991, Sødal 1989, Vatn 1989).

Denmark and Canada's Quebec Province have implemented packages of state supports for alternative agriculture, including an on-farm research component, special farmer training and extension programs, financial assistance for conversion to chemical-free production, organic certification, and market development and promotional activities (Dubgaard and Sørensen 1988, Dubgaard et al. 1990, MacRae 1989, MacRae et al. 1990, Østergaard 1989).[6]

Riding the Coattails of Supply Control

By the mid-1980s, it was generally understood that "measures that create incentives to increase intensity in agricultural production also lead to stress on the natural environment" (and, we would add, on farm animals). There was also "a growing public perception that it is not logical to damage the environment to produce surpluses" (OECD 1989b:22). Like Sweden, many other nations adopted measures intended to control supply while also promoting resource conservation and environmental protection. Familiar instruments such as acreage set-asides, milk quotas, and dairy herd buyouts are clearly symptomatic responses, not cures for the underlying incentive distortions or the production-stimulating trajectory of new technology. These instru-

[6] The Danish organic farming program most resembles Sweden's and influenced Swedish policy. The Danish approach, however, is more closely tied to an economic strategy. Legislation in 1987 promoting organic farming put special emphasis on developing domestic and export markets. The product-certification standard was based on criteria laid down by the International Federation of Organic Agriculture Movements (IFOAM), and government inspection of organic farms and food distributors soon followed. An official organic products label was developed in consultation with farmers. And a subsidy of up to 40 percent was offered to producer cooperatives and food distribution enterprises for marketing ventures. Certification, inspection, labeling, and market development are all elements of a strategy to position Denmark as a major player in the fast-growing European market for chemical-free food. As an EC member with a green image, Denmark is in a unique position to capitalize on this opportunity (Dubgaard et al. 1990, Østergaard 1989).

ments, and many of the green initiatives described below, should thus be viewed as temporary: if and when excess production and depressed international prices disappear, it is doubtful whether such policies would be kept in place, or at least whether they would be financed at present levels. As we pointed out in the Swedish case, they would make less sense in either economic or ecological terms.

We focus here on the European Community and United States. Although the CAP's acreage-reduction policy has no formal cross-compliance requirement, grants are made to member nations to help finance extensification (1989) and land management agreements in environmentally sensitive areas (1985).[7] These programs cause slight reductions in aggregate input intensity and in acreage of surplus commodities. Germany makes the most extensive use of compensated input reduction to cut output; and in Britain, the Countryside Commission pays a premium above the normal set-aside payment to farmers who use environmentally preferred practices.

Skeptics contend that since these structural measures help to keep price supports in place, most of the economic benefit and even some of the environmental benefit is lost via indirect effects. Lower input intensity and production on enrolled land tend to be offset by higher intensity and yields on the remaining land. This defect reflects a general problem first demonstrated by Jan Tinbergen and long accepted by economists: a single policy instrument cannot attain multiple goals efficiently (Ervin and Tobey 1990, Harvey and Whitby 1988, Haxsen 1989, OECD 1989b).

On the whole, supply-control legislation is less confused with environmental and regional policy in the United States than in the EC. Nonetheless, the largest and arguably most comprehensive effort to tie supply control to resource conservation has taken shape in the United States. (Note that unlike the case in Europe, where biological diversity and rural social vitality are prime motives behind green initiatives, it seems easier in the United States to generate political consensus around resource conservation.)

The 1985 Farm Security Act's "Conservation Title" contained both carrots and sticks to link economics with ecology. First, the crop-yield indexes used to calculate income deficiency payments to program crop producers were frozen. This weakened the producer's incentive to intensify input use and raise yields as a way to secure larger federal

[7] The extensification directive requires a 20 percent reduction in per-hectare input intensity.

payments. Second, voluntary cross-compliance requirements were added to the deficiency payment program. Wetlands and erosion-prone land newly brought into production of program crops were made ineligible for payments (these are the so-called swampbuster and sodbuster provisions). Further, a producer of program crops on highly erodible land must have an approved soil-conservation plan by 1990 to remain eligible. We observe that these provisions do not make polluters pay, that farmers need not follow the conditions if they give up deficiency payments, and that enforcement is said to be lax (Reichelderfer 1990).

The 1985 act's Conservation Reserve Program (CRP) designated several regions with serious erosion problems and authorized that up to 16 million hectares of program crops be put into "long-term retirement" via planting of grass ley, trees, and so on. Again, participation is voluntary. Producers apply to enroll land, indicating the level of rent they require to forego planting cash crops. The Department of Agriculture sets the maximum rent it will pay in each region but does not publicize information about its "reservation price."[8] When the department and the farmer reach a mutually acceptable rent, a ten-year management contract is drawn up stipulating how the land shall be used.

CRP was hailed as an environmental success in its early years. As of 1989, 12 million hectares were enrolled, and by one estimate aggregate U.S. soil erosion was cut by one-third (or 609 million tons/year). The estimated benefits (in lower deficiency payments, less nutrient loss, and reduced off-farm pollution costs) were far greater than the program's $1.5 billion budgetary cost (Brown and Young 1990, Hinkle 1989). Skeptics contend, however, that the CRP "reached neither the potential soil erosion reduction nor the potential supply control benefits that could have been achieved at equivalent public cost," because of defects in the designation of regions, eligibility criteria, and rent setting (Reichelderfer 1990:219). A glaring example of omission is the Chesapeake Bay watershed: though its aquatic ecosystems and water quality are severely degraded, it was excluded from the 1985 CRP. Furthermore, an analysis of marginal costs and benefits casts doubt on whether expansion to the mandated 16 million hectares is socially justified (Aly et al. 1989, Batie 1990).

The 1990 Farm Act authorized several new voluntary conservation measures, extending and improving on the 1985 legislation. The CRP

[8] Evidently, it takes farmers only a few enrollment periods to form quite accurate notions of the rental ceilings in their regions, and bids quickly converge toward the ceilings (Schoemaker 1989).

added protection of wetlands and water quality to its eligibility criteria. In the Water Quality Protection Program, farmers in regions with dangerous levels of agrichemical residues in groundwater may receive up to $3,500 per year (per farm) if they contract to cut back fertilizer or biocide use for five years. The target enrollment is 4 million hectares by 1995. Under the Wetland Easements Program, farmers in designated areas are eligible if they commit to preserve wetlands for thirty years. The target enrollment is 0.4 million hectares by 1995.

In addition, an Integrated Farm Management Program Option (IFMPO) allows participants to reduce their economic dependence on program crops by generating some revenue from acreage enrolled in the set-aside. As in the EC's extensification scheme, they may grow cash crops using reduced chemical methods or grow soil- and nutrient-conserving crops such as sod and legumes on up to 20 percent of their program crop base acres. Where commercial sales from set-aside land had previously been prohibited, any nonprogram crops may now be marketed. Up to 10 million hectares may be enrolled in IFMPO.

These incremental improvements are a good illustration of the potential for fine-tuning. But they also reveal the vulnerability of green measures to changing priorities and budgetary pressures. When U.S. grain reserves fell and the federal budget deficit soared in the early 1990s, funding for the conservation provisions just described was cut to a fraction of the level authorized by the 1990 Farm Act (AFT 1991, IAA 1991a).

Green Agricultural Research: The Quest for Technological Fixes

Changes in agricultural research priorities prompted by greening tend to confirm economists' perception that technological innovation is largely endogenous, induced by market signals and policy signals. Changing research priorities in the public sector seem to have three main sources: first, a realization that past biases in agricultural technology have worsened pollution, resource depletion, landscape and biotope degradation, and animal health problems; second, gaps in scientific and technical knowledge that have delayed or undercut the effectiveness of green measures; and third, the conviction that more appropriate and powerful technological tools can improve both agriculture's economic and its environmental performance. This last notion is the "technological fix."

At the preventive level, there has been an emphasis on altering institutions responsible for research design and oversight: "Administrative arrangements are also being changed to ensure that environmental questions are given full consideration in agricultural research. In many countries, it is now common practice to appoint people with environmental interests to boards and committees allocating funds for research" (OECD 1989b:26).

More affirmatively, national research establishments have launched new projects in whole farm systems analysis, such as organic farm management, as well as the familiar single practice research, for example, on animal waste composting. Preexisting projects in such areas as botanical pest controls and mechanical weed control have been augmented and modified. The 1989 OECD survey presents a lengthy catalog of such initiatives (1989b:124–126). In most countries, green research findings are being incorporated into curricula at agricultural universities and extension training centers and are making their way into extension literature and field advising. In addition to Sweden, a few nations (Austria, Finland, Norway) finance sustainable agriculture research from taxes on fertilizer and agricultural chemicals (OECD 1989b, Young 1988).

This description must be tempered by a recognition that traditional productionist projects still capture a larger share of applied research funding than does alternative agriculture, even by a generous definition of "alternative." So long as farm price policy continues to reward input intensity and high yields, this bias is not likely to disappear. Furthermore, in times of fiscal austerity, green research may be more vulnerable to budget cutting because its support lobby is less firmly entrenched. The U.S. Congress, for example, originally authorized a $40 million annual budget for Sustainable Agriculture Research and Extension (SARE) in the 1990 Farm Act. This budget would have involved an increase from about 1 to 5 percent of federal agricultural research funding. When the budget cutting knives were put to work, however, the SARE appropriation was slashed by 85 percent, while more traditional research suffered proportionally small cuts (IAA 1991a).

There is a palpable intellectual, ideological, and political tension between agroecology and biotechnology: the macroscopic and microscopic scientific-technical visions of sustainable agriculture. In principle, a blending of the two approaches is feasible and desirable, but within most national agricultural establishments, the hegemonic view seems to be that molecular and submolecular biology, not agroecology, should be—and perforce will be—the dominant science in reshaping twenty-first-century agriculture. There is widespread faith that biotechnology holds the magic bullet promise of bringing economically efficient solutions to a wide range of agroenvironmental problems, especially chemical pollutants and human health hazards (Goodman et al. 1987, Molnar and Kinnucan 1989).

Pharmaceutical and chemical corporations have staked a multibillion dollar bet on this outcome, and their commitment tends to make it a self-fulfilling prophecy. In Reichelderfer's words, "Advances in biotechnology hold particular promise for the profitable development of products that maintain productivity while reducing the environmental consequences of agricultural production" (1990:226).

The stress on profitability raises the question, Who will profit? Since agrobiotechnology research is currently concentrated in the laboratories of global corporations, it can be assumed that biotechnology-as-commodity will eventually be profitable to them. Since demand for food is inelastic, it is likely that consumers will also benefit via lower food prices or slower price inflation. The overall thrust of high-tech inputs—such as the controversial bovine growth hormone bST and herbicide-resistant crop strains—seems to be toward output growth and further industrialization of agriculture. These tendencies have destructive implications for farm structure and rural communities, not to mention their problematic environmental and animal health effects. Farmers will continue to run on a technology treadmill. As long as productionist farm policies are in place, it is difficult to imagine a much different technological future (Buttel and Geisler 1989, Comstock 1988, IAA 1991b, Lacy and Busch 1989, Reichelderfer 1990).

10.2 Sweden as Leader and Follower

Swedish greening is well in advance of that of most other nations in several specific areas, including quantitative reduction in agricultural pesticides; financial, research, and extension support for alternative agriculture; prototype development of renewable agricultural bioenergy; safe recycling of municipal wastes on farmland; and protection of farm animals' well-being. Taken together, Swedish measures in the four realms emphasized in this book—reduction of pollutants, landscape preservation and nature conservation, sustainable agriculture, and animal rights—appear to be more comprehensive than those of other nations. Nonetheless, other governments pioneered many of the innovative green measures and influenced Swedish policy in the process.

Dutch and Danish legislative packages to reduce nitrogen leaching and denitrification predate Sweden's by one to two years. They introduced many of the instruments later adopted by Sweden, including stocking density limits, manure storage standards, seasonal spreading restrictions, and cover crop requirements.[9] According to Hurst (1991), Dutch and Danish measures to reduce agricultural pesticide intensity are also quite similar to Sweden's. Protection of farm landscapes for their biological richness, historical significance, and recreational value is a venerable tradition in Austria, Great Britain, the Netherlands, and Switzerland. The United States's Conservation Reserve Program is an innovative way to couple resource conservation with supply control. Policies to support alternative agriculture via research, extension, and marketing services can be traced back at least to the early 1980s in France and Denmark as well as in several Canadian provinces and U.S. states. And Switzerland's 1985 regulation requiring more humane treatment of laying hens was a landmark response to effects of factory farming on animals' well-being. (These examples are taken from Cloke 1988b, Dubgaard and Nielsen 1989, Ervin and Tobey 1990, Hanley 1991, OECD 1989b, and Young 1989.)

The objective conditions that have catalyzed green policy responses differ from continent to continent, nation to nation, and even locale to locale. Perhaps Sweden's leading role is most noteworthy because the country is not pressed by objective agroenvironmental problems as severe as those facing many other nations. Water and air pollution from animal wastes is far more serious under factory farming conditions in densely populated Holland. Dutch groundwater is tainted by residues from pesticides applied at eight times the Swedish rate on a per-hectare basis. Aquatic eutrophication is more urgent in Denmark, where 91 kilograms/hectare of nitrate leaching is nearly five times the Swedish level. United States's annual topsoil loss of 2 billion tons causes incomparably greater off-farm pollution and long-term nutrient depletion. Maintaining traditional agrarian landscapes has far greater commercial economic importance in tourist-dependent Austria and Switzerland.[10] And the social opportunity cost of land conversion to

[9] The Dutch livestock density provision that allows operators to spread excess manure on neighboring land under contract was imitated by Sweden in its leaching-control program for four southern counties.

[10] Both the charm and the ecological stability of Swiss subalpine agrarian landscapes are under siege from phenomenal tourist and recreational growth, combined with massive hydrological and highway expansion projects. In 1972, for example, the village of Leysin in southwest Switzerland had twenty-five operating farms; only four remain today. Farm youth have left to work in the tourist industry, and leisure chalets have sprouted up where brown

nonfarm uses is greater in Japan, where arable land per capita is less than one-fifth of Sweden's and where, some contend, the spread of golf courses poses a significant threat to the future of agriculture (Haxsen 1990, Hebbert 1988, Hurst 1991, McCormack 1990, Tamminga and Wijnands 1991).[11]

If it is true that agricultural pollution and depletion of farm resources are less critical in Sweden than in much of the capitalist world, the question naturally arises, Why has Sweden become a world leader in greening agriculture? Two keys are Swedish political culture and the institutions of the negotiated economy. Swedish citizens became highly sensitized to local and global environmental degradation in the 1980s, under the combined influence of public school curricula, mass media scrutiny, and sophisticated campaigns by environmental and animal-rights organizations. In cross-country surveys, Swedes give higher priority to environmental protection than do citizens of other industrial nations.[12] We have argued that these widespread public sentiments have influenced the behavior of the Swedish Farmers' Federation and all the political parties via two distinct pathways: ideological and strategic. Ideologically, political actors are ethical beings and citizens subject to moral suasion; strategically, they are opportunistic agents who find it expedient to respond to, and even ride the wave of, green public opinion.

Conceivably, Swedish farmers have stronger intrinsic convictions about their responsibility for the common good than do their peers in societies with more individualistic ideologies and more confrontational politics. One thinks of Great Britain and the United States. But it is also clear that the LRF, under Bo Dockered's astute leadership, has understood and even taken advantage of green public sentiments. This

---

Swiss cows used to graze. The disappearance of cattle, in particular, threatens biological diversity and soil stability on steeply sloped alpine meadows (Simons 1992).

[11] Japanese golf courses also threaten urban water quality because of their exceptionally heavy applications of herbicides, insecticides, and artificial coloring agents (McCormack 1990:9).

[12] In a 1989 Gallup poll conducted in Sweden, Great Britain, and the United States, for example, environmental protection was rated the top priority policy concern by Swedes, while it ranked fourth in the other two countries—behind socioeconomic conditions, crime, and drugs (SIFO 1989b).

Citizens of Nordic countries are considered to be very environmentally conscious. In a study of attitudes regarding materialist and postmaterialist values, 88 percent of Swedish respondents ranked "protection of nature from pollution" first or second among twelve issues. The corresponding percentages in other countries were Norway, 78 percent; Finland, 80 percent; Iceland, 81 percent; and Denmark, 85 percent. As noted in the text, Denmark has more serious pollution problems than Sweden to prompt its citizens' concern (Knutsen 1989:229).

is reflected in LRF's boycott of sewage sludge deposition on farmland and its proactive stance on pesticide reduction, cover crop requirements, and development of renewable bioenergy. LRF also recognized the potential for cashing in on Sweden's clean, green image by exporting organic and reduced-chemical food. This would extricate at least some Swedish farmers from the cost-price squeeze imposed by the new food policy. Our impression is that, for whatever mix of reasons, Swedish farm organizations moved more rapidly and decisively to catch the green wave than their counterparts in North America, Oceania, and most of Europe.

10.3 Convergent Trajectories?

The industrial nations are still at a relatively early stage in designing instruments to internalize agriculture's positive and negative externalities. None have taken the qualitative leap beyond piecemeal greening, to articulate a farsighted and holistic national vision of sustainable agriculture or to translate such a vision into policy. We are impressed, however, by how much learning has occurred since early 1980s: learning by doing, learning via international communication of scientific and technical knowledge, and learning by borrowing insights from others' policy experiments. Sweden may stand out from the crowd in terms of the overall pace and extent of its greening, yet its policymakers have borrowed elements of the Netherlands' manure management strategy, Britain's ways of protecting scenic and historic landscapes, Denmark's alternative agriculture program, and Switzerland's poultry-housing regulations.

As knowledge about which green measures are effective and which are ineffective spreads internationally, further convergence toward the most effective policy concepts is very likely. Yet nations will continue to differ in many ways: in the nature and severity of their agroenvironmental problems, the economic importance of their farm and food sectors, the political clout of agroindustrial lobbies vis-à-vis environmental and alternative agriculture movements, and the strength of their citizens' green convictions and willingness to pay for them. These differences militate against policy convergence. So does the need for localized fine-tuning of green instruments such as input taxes, land-management contracts, and cover crop requirements introduced in the 1980s.

Nonetheless, there is certain to be a powerful political pressure for—and heated contention over—international harmonization of agroenvironmental standards and policies. In Europe, the farm policy gap between the EC and the European Free Trade Area countries was so wide that the negotiations to create a nineteen-nation European Economic Space in 1991 deliberately excluded agriculture. In the 1990s, however, the European Community is poised to add several new members, and agricultural and environmental policies will surely be on the bargaining table. New members will have to conform to the EC's growing body of common environmental standards and policies. But as they enter the web of intra-EC alliances, they will also gain some leverage over the way such standards and policies evolve.

A central theme in our analysis has been that getting farm input and output prices right is a precondition to achieving real consistency between economic and environmental objectives. The international pressure to reduce domestic market distortions comes largely from nations seeking to liberalize farm commodity trade. The pressure will not disappear, whatever the results of the Uruguay Round. Indeed, if and when economically motivated trade restrictions are rolled back, restrictions based on environmental justifications seem certain to become more contentious.

In nation after nation, momentum for a greener agricultural policy accelerated in the latter 1980s. But it cannot be assumed that this trend will continue in the 1990s. That depends on the answers to a series of key questions. If the growing prosperity of the 1980s gives way to a lengthy period of economic stagnation, as many economists predict, will a renewed preoccupation with bread-and-butter problems erode green priorities? Will artificially inflated farm prices, so closely linked to pollution, distorted land use, and industrialized livestock production, finally give way under fiscal pressure? Would a market-oriented agriculture strengthen or weaken the commitment to green policy measures, many of which were rationalized as supply-control measures? Will biological and chemical fixes for environmental problems undermine public support for alternative agriculture? Will environmental agencies encroach further into the traditional territory of agricultural administrations? Will agroenvironmental policy's bias toward a mix of voluntary and command-and-control measures persist, or will the use of economic levers expand? These are among the unknowns that face Sweden and other advanced industrial nations in the waning years of the twentieth century.

11

Globally Sustainable Agriculture in a New Millennium: Swedish Lessons

For much of the 1970s, the industrial nations' agricultural policies were shaped in reaction to a perceived world food crisis. Reactive productionist policies contributed to the unfolding farm crisis of the 1980s. As we have seen, economic crisis symptoms included downward pressure on farm incomes and asset values, chronic excess production, depressed and unstable international prices, and increasingly costly public support measures. Farm policy was deeply implicated as a cause of these problems.

In the 1990s, agricultural policy throughout the advanced capitalist world is being reshaped by two sets of values, pressures, and problems: two updated versions of the agrarian question. One is grounded in an economistic paradigm and the other in a sustainability paradigm.

The dominance of economism in current agricultural policy is reflected not only in Sweden's 1990 food policy, but also in the European Community's May 1992 CAP reform plan, the nearly completed Uruguay Round trade agreement, and a host of market-oriented measures adopted by other industrial countries.[1] Whatever political obstacles and countertrends may arise, the long-term trajectory of economistic agricultural policy is toward a more thoroughly industrialized and globally integrated system of agricultural production, food processing, and distribution. At the farm level, intensified competitive pressure, along with ceaseless technological developments and continuing vertical integration into agroindustrial complexes, seems certain to rein-

[1] Jonathan Rauch (1991), for example, notes that Austria has eliminated its food export subsidies, Norway and Switzerland are decoupling farm income maintenance from commodity pricing, and Japan has cut its support price for rice four times since 1985.

force past trends toward larger farm scale, increased specialization in the production of bulk commodities, and concentration of farm capital. In this unfolding scenario, the future economic significance of small farms, diversified operations, and high-value niche-market production is problematic. The policymakers who are dismantling the state's regulatory apparatus and substituting market-oriented measures, at both the national and international level, are certainly not oblivious of these implications. For the time being at least, neoliberalism is in the ascendancy and "capital is resocializing the state" (McMichael and Myhre 1990).

Economistic norms—efficiency, competition, free choice—coexist uneasily with the green ideology that underpins the sustainable agriculture movement. As discussed in this book, the sustainability vision actually combines two perspectives, one local and historically rooted, one global and future-directed. A core commitment at the local level is to preserve what remains of the biological richness, cultural traditions, and amenity values associated with preindustrial agriculture. Local sustainability is thus as much a social and cultural as an ecological concept. Green convictions of this type seem to be most widespread in the advanced industrial nations. Viewed through Third World eyes, they may seem to be a luxury affordable only in affluent societies, where food supply is secure and sustenance absorbs only a small share of resources and personal income.

Global sustainability is about conserving and maintaining the resources—and institutions—needed to ensure food security in an uncertain and risky world. (Parenthetically, economic efficiency, based on a thorough accounting of long-term and nonmarket costs and benefits, is a necessary condition for sustainability.) Among the major concerns that motivate this sustainability commitment are population momentum; depletion of and competing demands for land, water, and energy resources; the adverse environmental impact of industrialized agriculture in advanced capitalist and former socialist nations, and increasingly in the Third World as well; and signs that human-caused disturbance of the global atmosphere and climate may profoundly disrupt future agriculture.[2] If we consider hunger to be the ultimate indication of a nonsustainable food system, then it must always be kept in mind that food security is a matter of distributional entitlements as well as adequate production.

[2] We note that most simulations of global climate change indicate a beneficial overall effect on north European agriculture. In the Swedish case, however, beneficial temperature and

In practice, the economistic and sustainable agriculture visions are opposed on many points. Here we explore measures to mediate the differences: to find common ground, using Sweden as an example.

Sweden is a minor actor on the world stage, and agriculture has never been one of its well-known roles. The question thus naturally arises, What can the rest of the world learn, either affirmatively or negatively, from Sweden's recent agricultural policy experience? In a play, the character actor with a small role compels the audience's attention by personifying distinctive qualities and sharpening the focus on central themes in the unfolding plot. Sweden might play such a role in an unfolding drama of the 1990s: the search for principles and practices to guide the transition to an economically efficient and sustainable agriculture. Pierre Crosson (1992:14) reminds us that "the spatial scale for discussions of sustainable agriculture is global," so what happens in Sweden, with its mere one hundred thousand farmers and three million hectares of arable land, will be of international interest only to the extent that it contains lessons with broader applicability. What, then, can be learned from a small, chilly country with an open economy, a strong emotional and hedonistic attachment to the countryside, and a highly developed environmental consciousness?

In pragmatic terms, several Swedish environmental and economic initiatives might serve as affirmative or cautionary lessons to other industrial nations, West and East. Parenthetically, there is a strong case that both the greening process and economic deregulation of agriculture in the industrial North will benefit the South, via lower demand for nonrenewable resources such as fossil fuel and rock phosphate, reduced greenhouse gas emissions, advances in agroecological knowledge, and improved market conditions for staple food exporters (French 1990).[3] At a more philosophic level, we reflect on the possible lessons to be derived from Sweden's halting, piecemeal, conflict-ridden, and unfinished evolution toward a collective vision of sustainable agriculture. Sweden, together with its environmentally friendly neighbors,

precipitation changes could be offset by sea level rise, increasing pest damage, and deterioration of the boreal forest ecosystem.

[3] The results of past technology transfers to Third World agriculture caution us against excessive optimism about a greener Green Revolution in the future. Nonetheless, judicious use of advances in basic and applied agroecological knowledge, which is a core component of greening in the industrial nations, could facilitate the development of more sustainable technical and managerial systems in Third World agriculture.

has a unique opportunity to stimulate the universal greening of agriculture, by persuasion and by example.

11.1 Practical Lessons from Sweden: Process, Goal Structure, and Instruments

Political Process

The economic pressure for a sweeping agricultural policy reform in the latter 1980s grew out of Sweden's macroeconomic problems (inflation, budget deficits, labor shortage), and the widespread support for green agricultural initiatives reflected a broad politicization of the environment. Perhaps the first lesson is that far-reaching changes in farm policy require an atmosphere of crisis around societal problems of broad scope, to which agriculture contributes conspicuously (even if not as the most important cause).

Several years of debate about the purpose of agricultural policy and revelations about the existing policy's failure to fulfill its goals led up to the 1990 reform. Perhaps a second lesson is that one precondition for decisive change in a long-established policy regime is lengthy, open, and relatively enlightened public discourse. The Swedish debate, which was played out prominently in the media, revealed several things: the inefficiencies and inequities of farm price supports, the vested interests served by the old regime, the regime's high cost to consumers and taxpayers, and its environmental contradictions. In the end, even pro-farmer political parties could not afford to stand outside the consensus favoring reform.

Two Swedish political institutions have been especially important in the greening of agricultural policy and might serve as examples to other nations: first, the expert commissions assigned to translate broadly stated objectives into detailed policy options, giving them a stamp of nonpartisan legitimacy; and second, the remiss process, giving all concerned groups and individuals a voice in legislative debates. Both institutions countervail against the power of narrow economic interests and bureaucratic habits of behavior.

Two sets of actors, the mass media and environmental organizations, played crucial roles in greening the Swedish agricultural policy debate. Environmental activists elsewhere might learn from the way Swedish environmentalists have successfully positioned themselves as respected experts and as representatives of the public interest and from the way nature conservationists, historic preservationists, members of the alternative agriculture movement, and animal-rights activists joined forces, took advantage of media opportunities, and entered the parliamentary debate to move their causes to the center of the political agenda.

In sum, it is possible to dismantle a defunct policy regime if it is done cleverly and in a socially acceptable manner. Reform-minded politicians, while scathingly critical of the Swedish food system, carefully avoided blaming farmers for the mess. This civility fostered an atmosphere that allowed the farm lobby to accept income compensation and conversion assistance. Indeed, the multistage legislative process gave the LRF a chance to save face with its members by winning concessions and with the public by showing an environmentally friendly profile. The other side of this story, of course, is that farmers, with allies in agroindustry and in the Center and Moderate parties, were still able to muster considerable political clout. The concessions that marked the path to a consensus food policy increased its cost and reduced its goal attainment.

Goal Structure

As we have seen, a gradual shift in official agricultural priorities began in the early 1980s, paving the way for the 1990 reform. One lesson from this experience may be that reform-minded leaders must work patiently to change public perceptions by increments before they can fundamentally challenge the status quo. National food security is a case in point. In the 1990 legislation, food security retains high priority. But after several years of public discussion, its meaning has been altered to fit an international environment in which peacetime food sufficiency has dubious insurance value and import-dependent farm technology is actually a source of vulnerability. The national security rationale for high domestic farm prices and import barriers has been discredited.

By 1985, the commitment to farm income parity was already eroding. In part, this trend was a by-product of new thinking about national security needs. In part, it was a revival of old economistic arguments: that policies in a labor-scarce economy should not discourage workers from exiting a declining sector, and that commodity price supports, capitalized into asset values, are an inefficient and ineffective way to achieve income goals. Rethinking income parity was also based on considerations of distributional equity, since farm price supports were in large measure a transfer of income from wage-earning consumers to land-rich farmers. Rapid food price inflation in the 1980s made it politically easier to tilt the balance from producer to consumer interests, and it is noteworthy that the Social Democratic government called for a new *food policy,* not another farm policy.

Finally, environmental protection was reinstated as an agricultural policy goal in 1984, though it took several more years to articulate detailed green objectives and put policy measures in place. As a tactic, giving priority to environmental goals was one more way to undermine the old policy regime, which had created perverse incentives regarding the farm landscape, input intensity, livestock density, and the treatment of farm animals.

Instruments and Timetables

Several general insights can be gleaned from Sweden's recent initiatives, but it should be kept in mind that Swedish policymakers have learned their lessons through trial and error. There is still plenty of room for improvement in policy design and execution.

Sweden has made progress in substituting more "transparent" targeted measures for blunt, cost-ineffective instruments such as price supports and milk quotas. The new policy includes directly targeted measures to encourage such public goods production as food security, landscape protection, and the economic vitality of sparsely populated regions. There is increased reliance on supply-side competition and consumer demand to direct farm resource allocation, reducing if not totally eliminating the old policy's bias toward intensive input use, monoculture, and excess output. It is important to note, however, that some potential economic benefits will be lost so long as high and uneven import tariffs remain in place.

There has been some progress toward implementing the Polluter Pays Principle and the analogous Producer Compensation Principle for public goods providers. Examples of the former are pesticide dosage taxes and cover crop requirements. A prime example of the latter is payment for maintaining biologically rich pastures and meadows. We hasten to add that some PPP measures, such as manure storage requirements, and some PCP measures, such as subsidies for converting land to energy crops, are not cost-effective as currently designed. And as noted earlier, most of the cost entailed by internalizing negative externalities is ultimately passed on to consumers because of low demand elasticities and continuing import protection.

Cost-effectiveness has at long last become a way of thinking that permeates all policy commissions, and Sweden is increasingly replacing command-and-control measures with economic incentives to protect the environment. Examples of more cost-effective instruments are the risk-based biocide dosage tax, the tax on carbon emissions as a stimulus to agricultural energy conservation and biofuels development, and the livestock density provision that allows farmers to increase stocking levels by contracting to spread excess manure on neighboring land. Economic levers are not foolproof, however, as illustrated by the defects of a uniform fertilizer tax.

In addition to growing emphasis on ecologically sustainable agriculture, there is great interest in agriculture's contribution to a more sustainable national economy. The two main contributions, one actual and one potential, are safe recycling of municipal sewage sludge onto farmland and cultivation of bioenergy crops. The use of sludge as a soil amendment is backed by several years of experience, and steady improvement in sewage-processing techniques has cut the risk of contamination from heavy metals and hazardous organic compounds.

Agricultural bioenergy is still at the prototype stage of development, but the present research commitment to plant genetics and botany, agronomy, mechanization, materials handling, and industrial processing is likely to have a high, long-run economic and environmental payoff. Apart from the quantitatively minor role of energy crops in reducing carbon emissions and increasing national energy security, they could also cut nitrogen leaching and even improve long-term food security by keeping arable land in good tilth. (These effects depend heavily on the cultivar and production system.) On the downside, large-scale monocultures of woody species such as willow and hybrid aspen would detract from the landscape's biological diversity and visual/recreational amenities. Reconversion to food production in a national emergency would be extremely costly.

In view of all these factors, and given the present state of knowledge, it would be a serious economic mistake to subsidize further land conversion or establish price supports for energy crops. A "correct" level of carbon emissions tax would create the appropriate incentives to develop biofuels on a commercial scale. Since Sweden is punishing itself by imposing a carbon tax while most other nations continue to free ride, what we earnestly hope is that an international agreement will be reached to tax carbon emissions or create international quotas and tradable permits.

The common Swedish practice of setting tangible goals with explicit timetables has both psychological and efficiency advantages. These are exemplified by the targets for manure management and humane animal husbandry. The lead-in times are short enough to pressure farmers to adjust their practices, but long enough to allow them to amortize existing capital and adjust efficiently through learning by doing. With three, five, or even ten years of lead time, hastily legislated instruments can be debugged and fine-tuned; research institutions can upgrade technology and management systems.

Many green initiatives reflect an awareness that complex pollution problems and diverse farm conditions require comprehensive but flexible policy packages. The American social critic H. L. Mencken quipped that for every complex social problem there is a solution that is simple, straightforward—and wrong. With some notable exceptions, Swedish green initiatives have not fallen into the trap of oversimplification. The programs for nitrogen leaching and biocide reduction are prime examples of the flexible package approach to problem solving. The agricultural bioenergy experiment, with its farmland conversion supports, agronomic and engineering research component, exemption from the carbon tax, and public investment in energy-generation prototypes, appears to be another package approach in the making, though one with serious bugs still in the system.

Food-security instruments have been adapted to post–Cold War realities, especially the reality of Swedish dependence on imported fossil fuels and farm inputs. As we have stressed, the new measures improve near-term cost-effectiveness. They should, however, be founded on more sophisticated risk analysis, especially regarding Sweden's indirect long-term vulnerability to global population pressure and possible ecological crises. The principal defect of the present policy, in our assessment, is that it encourages excessive afforestation of arable land—agriculture's prime renewable resource.

Finally, Sweden is slowly moving toward more effective and far-sighted ways to integrate agriculture into a comprehensive rural economic development strategy. The conception of sustainable agriculture now taking shape combines management contracts to maintain an open and varied landscape with new forms of economic opportunity for part-time farmers, especially high-value niche-market products, value-added enterprises, and off-farm employment. If agriculture is to survive in the forest and mixed farm-forest regions—for its social, cultural, and ecological values—such an integrated rural development strategy is imperative.

11.2 Lessons for Sweden

Learning is not a one-way street, and we have recorded numerous reciprocal influences on Sweden. Ideas about how to revise and even dismantle agricultural regulatory systems have been in the air for many years. In the 1980s, Swedish policymakers frequently referred to the pioneering efforts of Australia and New Zealand. As mentioned, other highly protectionist nations, as well as the European Community and the United States, are also becoming more market-oriented.

The Swedish injection of greater competition into the food processing and distribution industries owes much to comparisons with European Community policy. The EC has been able to maintain a high level of economic support for farmers at a comparatively low cost in consumer welfare and resource misallocation by encouraging competition in food processing and distribution.[4] (Of course, the EC has the advantage of a market more than forty times as large as Sweden's.) Sweden's 1990–91 tariff reduction on processed foods and its antitrust measures against agribusiness collusion were informed by EC practice.

The greening of agriculture too is a reciprocal learning process, and Sweden has taken important lessons from Dutch and Danish manure-management policies, Swiss animal-welfare legislation, and landscape-management agreements in several nations. There are many other sustainable agriculture initiatives around the world that Swedish policymakers are considering or should consider.

There is general agreement that both consumers and rural ecosystems would be healthier if food were produced with smaller amounts of fertilizer and biocides. Direct subsidies to organic farmers, however, as used in Sweden from 1989 to 1992 and in Germany and Denmark, are not a cost-effective way to encourage reduced-input agriculture. This observation is true a fortiori once the rationale of cutting excess production is gone.

The social payoff to supports for low-input and chemical-free farming would be higher if the market scope for its products were greater. The LRF and others are convinced that Sweden's "clean" soil and

[4] Much of the lower cost to consumers is, of course, offset by higher taxes to finance agricultural programs and export dumping.

low-chemical agriculture offer a great export opportunity for high-value foods in western Europe. It is clear that many Swedish farmers will not survive in freer international competition if they continue to produce generic bulk commodities. Consumers in industrial nations increasingly select and pay premium prices for food produced on clean soils and with reduced fertilizer, biocides, and growth regulators. But Sweden currently lacks a coherent strategy to differentiate, promote, and supply environmentally friendly food products at home, much less on the Continent. We believe much could be learned from California's business-government partnership to promote and market organic produce, a venture that has moved beyond the niche-market stage to hundreds of millions of dollars in annual sales.[5] Likewise, Denmark's comprehensive incentives to stimulate the production, processing, labeling, advertising, and export of organically grown food should be studied closely.

We have commented favorably on Sweden's growing research support for low-input and ecological agriculture, but there is an unnecessarily wide gap between the findings from sophisticated scientific experiments and practical recommendations for commercial farmers. There has been some increase in whole farm systems research, tied in with on-farm experiments that are replicated across a number of working farms. Yet Swedish research planners would do well to learn from the U.S. Department of Agriculture's SARE (Sustainable Agriculture Research and Education) program, which has moved considerably farther in this fruitful direction. Integral to the SARE approach is a dialogue between scientists and practicing farmers that has an important payoff: the scientists' better grasp of real-world technical, managerial, and economic predicaments, both those that arise from present chemical-intensive methods and those associated with conversion to ecologically sustainable systems (Lockeretz and Anderson 1992).

Finally, as the Swedish population becomes ever more concentrated in and around large population centers, diverse and creative approaches are needed to sustain some agriculture on the urban fringe and tie remaining farms into community life.[6] Among the many models worth exploring, we note two: the Dutch and Japanese methods of integrating agricultural land use directly into long-term town planning, and the grassroots movement called "community-supported agricul-

[5] Several other U.S. states, notably Texas and Massachusetts, also vigorously promote their reduced-chemical and organic produce in national and international markets.

[6] Swedish municipalities already have strong negative sanctions against major land-use changes. The emphasis here, however, is on affirmative measures to support farming within the web of communities.

ture" (CSA), which spins a web of financial, market, and cultural ties between residents and farmers. CSA projects can be found in a growing number of communities scattered across northern Europe and parts of North America.

11.3 Part of the Solution or Part of the Problem? Contributing to Globally Sustainable Agriculture

In conclusion, we return to the meaning of sustainable agriculture: it is multigenerational in perspective, global in scale, and sociocultural as well as economic and ecological in scope. Starting from this conception, we raise two final questions. Is Sweden on a path toward sustainability? If other industrial nations adopted policies similar to Sweden's, would the prospects for global sustainability be improved? Our response to both questions is a qualified yes.

In the 1970s and 1980s, the trajectory of Swedish agriculture was distinctly nonsustainable. The production system depended heavily on imported, nonrenewable inputs; it was a source of increasing air and water pollution; its excess production helped distort international markets; it drove small and medium-size family farms, the historic backbone of rural communities, into obsolescence; and it relied on inefficient supports to keep some vestigial farming alive in regions not blessed by mild climate and fertile soils.

Recent green initiatives and the 1990 food policy redirected Swedish agriculture toward sustainability in several ways:

- Agricultural pollutants, including greenhouse gases, have been reduced and will be cut still further as the new food policy comes fully into effect
- Nonrenewable resource dependence is on the decline while maintenance of soil fertility by regenerative means is on the increase
- A significant part of the urban waste stream is recycled safely onto agricultural land, where it adds nutrients and organic matter
- Renewable bioenergy research and commercial experiments show considerable promise
- The biological diversity and amenity values associated with pastures and meadowlands are to a considerable extent protected by targeted landscape payments
- Research expenditures on low-input and alternative agriculture have grown several fold, and environmental risk analysis is increasingly incorporated into research design

- Small and part-time farming is to be more effectively integrated into regional development strategy
- The elimination of excess production and export subsidies contributes in a small way to international price recovery and augments the foreign exchange earnings of food-exporting nations. Higher, more secure earnings could ease their transition to sustainable production systems, especially if importers require that commodities be produced in an environmentally friendly manner.

If these positive steps are carried further in future agricultural and environmental policies, the prognosis for sustainable agriculture will continue to improve.

If we look beyond the new food policy's five-year transition period, however, we have profound doubts about likely developments in commodity markets, agricultural technology, and the farm economy. If, as seems probable, a liberal food trade regime takes shape, "survival of the fittest" competition and global concentration of agroindustrial capital will be unleashed. In such an environment, it is difficult to imagine more than a handful of large bulk commodity producers in Sweden surviving beyond the current generation of farmers. Apart from niche-market farming, how could Sweden's comparatively small family farms possibly cope with the tremendous geographic advantages of Canadian wheat, Iowa pork, New Zealand butter, or Argentine beef?

The privatization of cutting-edge agricultural technology also makes the social and economic sustainability of farming as we know it problematic. If we extrapolate trends of the past quarter century, it is not difficult to imagine the nearly complete hegemony of an industrialized food system within one generation. Farm technology would flow from the biotechnology, chemistry, and engineering labs of global corporations, and producers on the ground would become thoroughly subordinated within giant agroindustrial networks (see Goodman et al. 1987, Lacy et al. 1988, Sagoff 1988).

Family farms and agrarian communities would be progressively marginalized, as the returns to labor and capital on old-fashioned farms fell so low that only hobby and niche-market operations survived. The critical mass of local agroindustrial activities, which reciprocally support and are supported by farming, would also disappear, taking agrarian communities with them. As factories in the fields became less and less distinguishable from other capitalist businesses, there would be an erosion of the social respect and self esteem that kept millions of families in farming for a half century following World War II, despite their arduous labor and limited economic rewards.

An agriculture that was environmentally sustainable but had few farms and no farm communities would indeed be a pyrrhic victory. Yet this result is logically entailed by the premise underlying Sweden's 1990 food policy: agriculture should be treated like any other economic sector.

Whether in the next generation dark spruce forest takes back most of the landscape opened to daylight by generations of past farmers depends largely on three conditions, only one of which is fully under Swedish control. The first is how much Swedish citizens are willing to pay to protect and nurture the green agriculture that is just now being born—or reborn. Our research suggests that a full public debate about the social costs and benefits of alternative futures for agriculture would reveal a much greater collective commitment to sustainable agriculture than is reflected in the present policy.

The second uncertainty is whether some form of farsighted collective control can be exerted over the development and diffusion of agroindustrial technology, to make modern scientific knowledge the servant of sustainable agriculture rather than its destroyer. The corporate commodification of technology and the rapid spread of new techniques across national borders make it clear that technology guidelines and their enforcement must be global to be effective. International technology policy is certain to be a major political-economic battleground in the new world order.

The third unresolved question is whether an international consensus can be reached on the meaning of agriculture: what legitimate values and moral commitments justify political intervention to override the Darwinism of competitive markets? Sweden and like-minded nations, indeed all friends of sustainable agriculture, must prepare for a long political contest over the rights of nations and communities to protect green values, across the entire spectrum of "happy pigs" to "smiling landscapes." We hope that this study will enlighten that contest.

References

The Swedish letters *å, ä,* and *ö* come at the end of the Swedish alphabet and are alphabetized here following *z*.

AERI (Agricultural Economics Research Institute). 1988. *Allmänhetens inställning till lantbruket mm* (The public's perspective on farming). Report no. 19-5939. Stockholm.

——. 1989. *Lantbrukets Struktur 1987* (Agricultural structure 1987). Stockholm.

——. 1992. *Allmänhetens inställning till lantbruket mm* (The public's perspective on farming). Report no. 19-1798. Stockholm.

AFT (American Farmland Trust). 1991. "The New Farm Bill Holds Promise for Conservation." *Northeast Farmland Update,* 2(3): 1–2.

Almaas, Reidar. 1991. "Farm Policies and Farmer Strategies: The Case of Norway." In W. Friedland, L. Busch, F. Buttel, A. Rudy, eds., *The New Political Economy of Agriculture.* Boulder, Colo.: Westview Press, pp. 275–288.

Aly, Hasan, J. Esseks, and Steven Kraft. 1989. "Conservation Programs of the 1985 Food Security Act." In Dubgaard and Nielsen, eds.: 85–97.

Anderson, Kym, and Yujiro Hayami. 1986. *The Political Economy of Agricultural Protection in East Asia in International Perspective.* Sydney: Allen and Unwin.

Andersson, Anna-Karin. 1985. *Vattenvårdsåtgärder i Ringsjöområdet: En företagsekonomisk analys* (Water conservation measures in the Ringsjö region: An enterprise economy analysis). Report no. 257. Uppsala: Department of Economics and Statistics, SLU.

Andersson, Rune. 1986. "Förluster av kväve och fosfor från åkermark i Sverige" (Losses of nitrogen and phosphorus from arable land in Sweden). Master's thesis, SLU, Uppsala.

Andersson, Åke. 1987. *Vårt jordbrukspolitiska system* (Our agricultural policy system). Report no. 284. Uppsala: Department of Economics and Statistics, SLU.

Andréasson-Gren, Ing-Marie. 1991. *Regional Management of Nitrogen Polluted*

Waters: Two Swedish Case Studies. Småskrift Series no. 42. Uppsala: Department of Economics, SLU.

Anton, Thomas. 1969. "Policy-making and Political Culture in Sweden." *Scandinavian Political Studies*, 4:88–102.

Arnold, Richard, and Claude Villain. 1990. *New Directions for European Agricultural Policy*. CEPS Paper no. 49. Brussels: Centre for European Policy Studies.

Aronsson, Kenneth, and Eva Pettersson. 1989. "Skatteutgifter i svenskt jordbruk" (Tax expenditures in Swedish agriculture). Degree essay no. 36. Uppsala: Department of Economics, SLU.

Arrow, Kenneth. 1974. *The Limits of Organization*. New York: Norton.

Axelsson, Svante. 1988. *NOLA och andra stödformer för öppethållande* (NOLA and other forms of support for maintenance of open land). Småskrift Series no. 15. Uppsala: Department of Economics, SLU.

Baldock, David. 1990. *Agriculture and Habitat Loss in Europe*. World Wildlife Fund International. CAP (Common Agricultural Policy) Discussion Paper no. 3. London: Institute for European Environmental Policy.

Batie, Sandra. 1990. "Agricultural Policy and Environmental Goals: Conflict or Compatibility." *Journal of Economic Issues*, 24(2): 565–573.

Belotti, Catherine. 1991. "Vad betyder mjölkkornas hälsa för ekonomin?" (What is the impact of milk cows' health on the economy?). *Fakta Ekonomi* (May). Uppsala: SLU.

Berg, Jan. 1993. *Swedish Public Opinion and the European Community*. Stockholm: Swedish Ministry of Foreign Affairs.

Bergman, Lars. 1989. "Tillväxt och miljö: En studie i målkonflikter" (Growth and environment: A study in goal conflicts). Appendix 9 to Lantbruks Utredning 90. Stockholm: Ministry of Finance.

Bergström, Hans. 1991. "Election Year 1991: Social Democracy in Crisis." *Current Sweden*, no. 381. Stockholm: The Swedish Institute.

Berlan, Jean-Pierre. 1990. "Capital Accumulation, Transformation of Agriculture, and the Agricultural Crisis: A Long-Term Perspective." In MacEwan and Tabb, eds.: 205–224.

Blandford, David, Harry de Gorter, Praven Dixit, and Stephen Magiera. 1990. "Agricultural Trade Liberalization: The Multilateral Stake in Policy Reform." In K. Allen, ed., *Agricultural Policies in a New Decade*. Washington, D.C.: Resources for the Future and National Planning Association, pp. 231–252.

Blandford, David, William Meyers, and Nancy Schwartz. 1988. "The Macro Economy and the Limits to U.S. Farm Policy." *Food Policy*, 13(2): 134–139.

Bolin, Olof, Per-Martin Meyerson, and Ingemar Ståhl. 1986. *The Political Economy of the Food Sector*. Stockholm: SNS Förlag.

Bolin, Olof, Ewa Rabinowicz, and Niels von Zweibergk. 1989. *Åkerfällan: Energiskog ingen lösning* (The farmland trap: Energy forest no solution). Stockholm: SNS Förlag.

Bonanno, Alessandro. 1989. "Changes, Crisis, and Restructuring in Western Europe: The New Dimensions of Agriculture." *Agriculture and Human Values*, 7(1 & 2): 2–10.

——. 1991. "From an Agrarian to an Environmental, Food, and Natural Resource Base for Agricultural Policy: Some Reflections on the Case of the EC." *Rural Sociology*, 56(4): 549–564.

Bradsher, Keith. 1991. "The Mixed Blessing of Empty Bins." *The New York Times,* 6 November: D1.
Bromley, Daniel. 1989. "Entitlements, Missing Markets, and Environmental Uncertainty." *Journal of Environmental Economics and Management,* 17:181–194.
Brorsson, Kjell-Åke. 1989. *Ekonomiska effekter av omställningsbidrag till alternativ odling* (Economic consequences of conversion subsidies for organic farming). Alternativ Odling, no. 2. Uppsala: SLU.
——. 1991. "Ekologiskt lantbruk: Ett ekonomiskt produktionssystem?" (Organic farming: An economic production system?). In *An environmentally friendly agriculture: Consequences for enterprises and society.* Info report no. 175. Uppsala: SLU, pp. 23–46.
Brown, Lester. 1991. "The New World Order." In L. Brown, C. Flavin, and S. Postel, eds., *State of the World 1991.* New York: Norton, pp. 3–20.
——. 1992. "World Grain Takes a Spill." *Worldwatch,* 5(3): 35–36.
Brown, Lester, and John Young. 1990. "Feeding the World in the Nineties." In L. Brown, C. Flavin, and S. Postel, eds. *State of the World 1990.* New York: Norton, pp. 3–20.
Brundin, Sven, and Lena Rodhe. 1990. *Ekonomisk analys för hanteringskedjor för stallgödsel* (Economic analysis of distribution chains for manure). Report no. 118. Uppsala: Swedish Institute of Agricultural Engineering.
Buttel, Frederick, 1989. "The U.S. Farm Crisis and the Restructuring of American Agriculture." In Goodman and Redclift, eds.: 46–83.
——. 1990. "The Political Agenda for Sustainable Agriculture: Short-, Medium-, and Long-term Considerations." Paper presented at the Conference on Sustainable Agriculture, Santa Cruz, Calif., June.
——. 1992. "Environmentalization: Origins, Processes, and Implications for Rural Social Change." *Rural Sociology,* 57(1): 1–27.
Buttel, Frederick, and Charles Geisler. 1989. "The Social Impacts of Bovine Somatotropin: Emerging Issues." In Molnar and Kinnucan, eds.: 103–120.
Byé, Pascal. 1989. "The Restructuring in the Agricultural Supply Sector and Its Consequences on Agricultural Production." *Agriculture and Human Values,* 7(1 & 2): 22–28.
Cahill, Carmel, and Gérard Viatte. 1993. "A Fallow Year for Agricultural Reform." *OECD Observer,* 182 (June/July). Paris. pp. 4–6.
Cline, William. 1992. *The Economics of Global Warming.* Washington, D.C.: Institute for International Economics.
Cloke, Paul. 1988a. "Land-use Planning in Rural Britain." In Cloke, ed.: 18–46.
——, ed. 1988b. *Rural Land-use Planning in Industrial Nations.* London: Unwin Hyman.
Cochrane, Willard. 1979. *The Development of American Agriculture.* Minneapolis: University of Minnesota Press.
Comstock, Guy. 1988. "The Case against bGH." *Agriculture and Human Values,* 5(3): 36–52.
Coulomb, Pierre, and Hélène Delorme. 1989. "French Agriculture and the Failure of Its European Strategy." In Goodman and Redclift, eds.: 84–112.
Crosson, Pierre. 1986. "Agricultural Development: Looking to the Future." In

W. Clark and R. Munn, eds., *Sustainable Development of the Biosphere.* London: Cambridge University Press, pp. 104–139.
——. 1989. "Greenhouse Warming and Climate Change." *Food Policy,* 14(2): 107–108.
——. 1992. "Sustainable Agriculture." *Resources,* 106:14–17. Washington, D.C.: Resources for the Future.
Crotty, James. 1990. "The Limits of Keynesian Macroeconomic Policy in the Age of the Global Marketplace." In MacEwan and Tabb, eds.: 82–100.
Csaki, Csaba. 1990. "Agricultural Reform in Eastern Europe in the 1990s." *American Journal of Agricultural Economics,* 72(5): 1233–1242.
Dagens Nyheter. 1992. "Halvhård EG-boll tillbaka" (Halfhard EC ball back to us). 1 August. Stockholm, p. A2.
Dockered, Bo. 1989. "En jordbrukspolitik för sekelskiftet" (An agricultural policy for the turn of the century). In *Svenskt Jordbruk inför 1990-talet.* Report no. 43. Stockholm: Royal Academy of Forestry and Agriculture, pp. 59–66.
Drake, Lars. 1987. *Värdet av bevarat jordbrukslandskap* (The value of preserving agricultural landscape). Report no. 289. Uppsala: Department of Economics and Statistics, SLU.
——. 1988. "Sätt pris på miljön: Jordbrukslandskapet värderat" (Put a price on the environment: Agricultural landscape valued). *Ekonomisk debatt.* 5:568–577.
——. 1989. "Swedish Agriculture at a Turning Point." *Agriculture and Human Values,* 7(1 & 2): 117–126.
——. 1990a. "Styrmedel" (Policy instruments). In Frank Petrini, ed., *Mot ett uthålligt och miljövänligt jordbruk.* Report no. 28. Uppsala: Department of Economics, SLU.
——. 1990b. *Ekonomisk analys av kväveefterfrågans priskänslighet* (Economic analysis of demand elasticity for nitrogen). Foundation Report. Stockholm: Commission on Economic Instruments in Environmental Policy.
——. 1993. *Relations among Environmental Effects and Their Implications for Efficiency of Policy Instruments.* Uppsala: Department of Economics. SLU.
Drake, Lars, Mari Andersson, and Karl-Ivar Kumm. 1991. *Har jordbruket i Rottnadalen någon framtid?* (Does farming in Rottnadalen have a future?). Småskrift Series no. 48. Uppsala: Department of Economics, SLU.
Drake, Lars, and Frank Petrini. 1985. "The Influence of the Agricultural Land Care Act and the Nature Conservation Act on Natural Resource Use in the Swedish Agricultural Sector." In F. Lechi, ed., *Agriculture and the Management of Natural Resources.* Kiel: Wissenschaftsverlag Vauk, pp. 29–41.
Dubgaard, Alex. 1990a. *Danish Policy Measures to Control Agricultural Impacts on the Environment.* Report no. 32. Copenhagen: State Agricultural Economics Institute.
——. 1990b. *The Danish Nitrate Policy.* Copenhagen: Department of Economics and Natural Resources, Royal Veterinary and Agricultural University.
Dubgaard, Alex, and A. H. Nielsen, eds. 1989. *Economic Effects of Environmental Regulation in Agriculture.* Kiel: Wissenschaftsverlag Vauk.
Dubgaard, Alex, Per Olson, and Søren Sørensen. 1990. *Ekonomien i ekologiskt jordbrug* (Economics in ecological agriculture). Report no. 54. Copenhagen: State Agricultural Economics Institute.

Dubgaard, Alex, and Søren Sørensen. 1988. *Ecological and Biodynamic Agriculture: A Statistical Investigation.* Report no. 43. Copenhagen: State Agricultural Economics Institute.

EC Commission. 1990. *Agriculture and Environment.* Brussels: Committee on Agricultural Problems, UN FAO Economic Commission for Europe. FAO/ECE/AGRI/WP.3/R.119.

Edman, Stefan. 1990. "Vad kostar våra blåklockor?" (What do our bluebells cost?). *Dagens Nyheter,* 8 June. Stockholm.

Elster, John. 1989. "Social Norms and Economic Theory." *Journal of Economic Perspectives,* 3(4): 99–117.

——. 1991. "Argumenter og forhandlinger" (Arguments and negotiations). Unpublished manuscript. Stockholm: Institute for Future Studies.

Eriksson, Marianne. 1989. *Ägarestrukturens förändringar inom privatskogsbruket i ett historiskt perspektiv* (Changes in ownership structure within private forestry in a historical perspective). Report no. 5. Uppsala: Department of Forest Industry and Market Studies, SLU.

Ervin, David, and James Tobey. 1990. "European Agriculture and Environmental Quality: Sorting through Incentives." Paper presented at the American Enterprise Institute Conference "Is Environmental Quality Good for Business?" Washington, D.C., June.

Esping-Andersen, Gøsta. 1985. *Politics against Markets: The Social Democratic Road to Power.* Princeton: Princeton University Press.

——. 1990. *The Three Worlds of Welfare Capitalism.* Princeton: Princeton University Press.

Etzioni, Amitai. 1988. *The Moral Dimension: Toward a New Economics.* New York: Free Press.

——. 1990. "A New Kind of Socioeconomics." *Challenge* (January–February): 31–32.

Eyerman, Ron. 1985. "Rationalizing Intellectuals: Sweden in the 1930s and 1940s." *Theory and Society,* 14:777–807.

Fahlbeck, Erik. 1989. *Samhällskostnader för jordbrukspolitiken: En nordisk jämförelse* (Welfare losses due to the agricultural policy: A Nordic comparison). Report no. 18. Uppsala: Department of Economics, SLU.

——. 1990. "Vinnare och förlorare i den nordiska jordbrukspolitiken" (Winners and losers in Nordic agricultural policy). *Fakta Ekonomi,* 2. Uppsala: Department of Economics, SLU.

Fogelfors, H., B. Johnsson, and O. Pettersson. 1992. *Miljövänlig bekämpning i jordbruket: En diskussion av möjligheter och konsekvenser* (Environmentally friendly biocide use in agriculture: A discussion of possibilities and consequences). Uppsala: Department of Economics, SLU.

Foreign Affairs Ministry. 1990. *Aktuellt i handelspolitiken* (Current issues in commercial policy). February 1990. Stockholm: Commerical Office.

Fornling, Pär. 1990. "Politikerna nöjda" (Politicians pleased). *ATL* 24(21) (25 May), p. 6.

Freeman, A. Myrick. 1993. *The Measurement of Environmental and Resource Values: Theory and Practice.* Washington, D.C.: Resources for the Future.

French, Hilary. 1990. *Green Revolutions: Environmental Reconstruction in East-*

ern Europe and the Soviet Union. Worldwatch Paper no. 99. Washington, D.C.: Worldwatch Institute.

Friedman, Harriet. 1991. "Changes in the International Division of Labor: Agri-food Complexes and Export Agriculture." In W. Friedland, L. Busch, F. Buttel, A. Rudy, eds., *The New Political Economy of Agriculture.* Boulder, Colo.: Westview Press, pp. 65–93.

Frykman, Jonas, and Orvar Löfgren. 1979. *Culture Builders.* New Brunswick, N.J.: Rutgers University Press.

GBD (Glesbygds Delegationen). 1984. *Glesbygden: Resurser och möjligheter* (Sparsely populated regions: Resources and possibilities). Ds I 1984:20. Stockholm: Ministry of Industry.

Gianessi, Leonard. 1989. "A Critique of the National Academy of Sciences Report *Alternative Agriculture.*" Unpublished manuscript. Washington, D.C.: Resources for the Future.

Gillberg, Björn. 1970. *Miljö, Ekonomi, Politik* (Environment, economy, politics). Stockholm: Wahlström and Widstrand.

Goodman, David. 1991. "Tendencies in the Reorganization of the Agri-food System." In W. Friedland, L. Busch, F. Buttel, A. Rudy, eds., *The New Political Economy of Agriculture.* Boulder, Colo.: Westview Press, pp. 37–64.

Goodman, David, and Michael Redclift, eds. 1989. *The International Farm Crisis.* New York: St. Martin's.

Goodman, David, B. Sorj, and J. Wilkenson. 1987. *From Farming to Biotechnology.* Oxford: Blackwell.

GP (Government Legislative Proposition). 1988. *Miljöpolitiken inför 90-talet* (The government's environmental bill). 1987/88:85. Stockholm.

——. 1990a. *Regeringens proposition 1989/90:50 om inkomstskatt för 1990, m.m.* (The government's bill on income tax policy). Stockholm.

——. 1990b. *Regeringens proposition 1989/90:146 om livsmedelpolitiken* (The government's bill on food policy). Stockholm.

——. 1990c. *En god livsmiljö* (The government's environmental bill). 1990/91: Bill number 90. Stockholm.

——. 1993. *Regerings proposition 1993/94: 3 om villkoren för vegetabilie produktionen efter budgetåret 1993/94* (Government's bill on conditions for vegetable production after budget year 1993/94). Stockholm.

Granstedt, Artur. 1990. *Fallstudier av kväveförsörjning i alternativ odling* (Case studies of nitrogen pollution in alternative farming). Alternativ Odling, no. 4. Uppsala: SLU.

Greenberg, Steven. 1989. "Sweden's Social Democrats Veer toward Free Market and Lower Taxes." *New York Times,* 27 October.

Grossman, Margaret. 1989. "Farmland and the Environment: Protection of Vulnerable Agricultural Areas in the Netherlands." *Agriculture and Human Values,* 7(1 & 2): 101–109.

Gulbrandsen, Odd, and Assar Lindbeck. 1966. *Jordbrukspolitikens mål och medel* (Agricultural policy's goals and means). Stockholm: Almquist and Wicksell.

——. 1969. *Jorbruksnäringens ekonomi* (The agricultural sector's economy). Stockholm: Almquist and Wicksell.

Haavelmo, Trgyve, and Stein Hansen. 1991. "On the Strategy of Trying to Reduce Economic Inequality by Expanding the Scale of Human Activity." In Robert

Goodland H. Daly, S. El Sarafy, B. von Droste, eds., *Environmentally Sustainable Economic Development: Beyond Brundtland*. Paris: UNESCO, pp. 41–50.

Hahn, Thomas. 1991. "Avloppsslam på åkermark eller deponi?" (Sewage sludge on arable land or landfill?"). Degree essay no. 61. Uppsala: Department of Economics, SLU.

Hanley, Nick. 1989. "Problems in Valuing Environmental Improvements Resulting from Agricultural Policy Changes: The Case of Nitrate Pollution." In Dubgaard and Nielsen, eds.: 117–132.

——. 1990. "An Update on the Institutional Situation Regarding Nitrate Pollution in the EEC." Unpublished manuscript. Department of Economics, University of Stirling, Scotland.

——, ed. 1991. *Farming and the Countryside*. Wallingford Oxfordshire, UK: CAB International.

Harvey, David. 1991. "Agriculture and the Environment: The Way Ahead." In Hanley, ed.: 275–321.

Harvey, David, and Martin Whitby. 1988. "Issues and Policies." In Whitby and Ollerenshaw, eds.: chap. 13.

Hasund, Knut Per. 1990a. "Åker (och) miljön åt skogen" [nontranslatable pun]. In Hasund, ed., *Styrmedel, livsmedel och livsmiljö*. Småskrift series no. 30. Uppsala: Department of Economics, SLU, pp. 39–48.

——, ed. 1990b. *Styrmedel, Livsmedel, Livsmiljö* (Policy instruments, food, environment). Småskrift Series no. 30. Uppsala: Department of Economics, SLU.

——. 1990c. *Styrmedel för kulturlandskapet* (Policy measures for the rural landscape). Småskrift Series no. 37. Uppsala: Department of Economics, SLU.

——. 1990d. *Styrmedel att begränsa växtnäringsläckaget* (Policy instruments to control nitrogen leaching). PM 91. Stockholm: Commission on Economic Instruments in Environmental Policy, Ministry of Environment and Energy.

——. 1990e. *Avgiftssystem att begränsa ammoniakemmissioner från stallgödsellagring* (Levy systems for reduced amonia emissions from manure). PM 95. Stockholm: Commission on Economic Instruments in Environmental Policy, Ministry of Environment and Energy.

——. 1991. *Landskapspolitiken i Sverige, 1960–1990: En empirisk utvärdering* (Swedish landscape policy, 1960–1990: An empirical evaluation). Report no. 41. Uppsala: Department of Economics, SLU.

——. 1992. *Den nye bonden och hans bygd: Producenten av landskapsvärden* (The new farmer and his countryside: Producer of landscape values). Agricultural Conference 1992. Public information report no. 177. Uppsala: SLU.

Hathaway, Dale. 1987. *Agriculture and the GATT: Re-writing the Rules*. Report no. 20. Washington, D.C.: Institute for International Economics.

Haxsen, Gerhard. 1989. "Decreased Intensity in Crop Production and Its Effects on Food Supply in Emergency Situations." In Dubgaard and Nielsen, eds.: 197–206.

——. 1990. "National Measures of Environmental Policy in Agricultural Production: Comparison of Denmark, Sweden, and the Federal Republic of Germany." Paper presented at the International Conference on Environmental Cooperation and Policy in the Single European Market, Venice, April.

Hebbert, Michael. 1988. "Rural Land-use Planning in Japan." In Cloke, ed.: 130–151.

Heckscher, Eli. 1915. *Världskrigets ekonomi* (The economy of the world war). Stockholm: Norstedt and Söners Förlag.

Hedlund, Stefan, and Mats Lundahl. 1985. *Beredskap eller protectionism?* (Preparedness or protectionism?). Malmö: Liber Förlag.

Hedvåg, Lennart. 1990. *Konventionella och alternativa system för avloppsvattenbehandling och slamanvändning* (Conventional and alternative systems for wastewater handling and sludge utilization). Småskrift Series no. 32. Uppsala: Department of Economics, SLU.

Hellstrand, Stefan. 1988. *Alternativ djurhållning: En behovsinventering* (Alternative animal husbandry: A needs inventory). Stockholm: Swedish Council for Forestry and Agricultural Research.

Hellström, Annette. 1991. "(LRF) Näringspolitiska Mål: Miljö" (LRF economic policy: Environment). Unpublished manuscript. Stockholm.

Hinkle, M. K. 1989. "Environmental Regulations in American Agriculture." In Dubgaard and Nielsen, eds.: 15–24.

Hirschman, Albert. 1971. *Exit, Voice, and Loyalty.* Cambridge, Mass.: Harvard University Press.

Hjelm, Lennart. 1991. *Jordbrukets strukturella och driftsmässiga utveckling* (Agriculture's structural and operations-related development). Stockholm: Länsförsäkringsbolagens Forskningsfond.

Hodge, Ian. 1991. "Incentive Policies and the Rural Environment." *Journal of Rural Studies,* 7(4): 373–384.

Holm, Helen, and Lars Drake. 1989. *Konsumenternas attityder till alternativt producerat kött* (Consumers' attitudes toward alternatively produced meat). Report no. 21. Uppsala: Department of Economics, SLU.

Holstein, Fredrik, and Niclas Åkesson. 1991. "Växande energi: En samhällsekonomisk analys av energi från åkermark" (Growing energy: A social-economic analysis of energy from arable land). Degree essay no. 76. Uppsala: Department of Economics, SLU.

Holusha, John. 1991. "States Lead on Environment and Industries Complain." *New York Times,* 1 April, p. D1.

Hurst, Peter. 1991. *Pesticide Reduction Programs in Denmark, the Netherlands, and Sweden.* London: World Wildlife Fund International.

Höök, Lars. 1992a. "Brygga över till EG" (Bridge over to the EC). *Land,* no. 36. (September), p. 4

——. 1992b. "Spannmålen rasar till 85 öre med ny EG-politik" (Small grains will fall to $0.14/kilo with new EG policy). *Land,* no. 22 (May), p. 4.

IAA (Institute for Alternative Agriculture). 1991a. "Sustainable Agriculture Programs Fare Poorly in House Appropriations." *Alternative Agriculture News,* 9(7): 1.

——. 1991b. "Final Agricultural Bill Gives LISA No Increase in Funding." *Alternative Agriculture News,* 9(11): 1.

——. 1991c. "Biotechnology May Change Agricultural Way of Life." *Alternative Agriculture News,* 9(11): 1.

Ihse, Margareta, Per Kågeson, Lars-Erik Liljelund, and Anders Wijkman. 1990. "Mycket prat, lite handling" (Much talk, little action). *Dagens Nyheter,* 5 June. Stockholm.

Jackson, Wes. 1989. "One Victory Is Not Enough." *The Land Report* (Fall): 18–24. Salina, Kan.

Jamison, Andrew. 1987. "The Making of the New Environmental Movement in Sweden." Unpublished manuscript. University of Lund Research Policy Institute.

Jensen, Per. 1990. "Djurskyddslagen fuskas bort" (The animal protection law is cheated away). *Dagens Nyheter,* 7 December. Stockholm.

Jessop, Bob, Klaus Nielsen, and Ove Pedersen. 1991. "Structural Competitiveness and Strategic Capacities: Rethinking State and International Capital." In Sven Erik Sjöstrand, ed., *Institutional Change: Theory and Empirical Findings.* Armonk, N.Y.: Sharpe, pp. 227–262.

Johansson, Per-Olof. 1987. *The Economic Theory and Measurement of Environmental Benefits.* Cambridge: Cambridge University Press.

Johansson, Valter. 1984. *Konsekvenser av ändrad resursanvändning i jordbruksproduktionen* (Consequences of changed resource use in agricultural production). Småskrift Series no. 9. Uppsala: Department of Economics, SLU.

——. 1992. "Vad kostar det att bibehålla åkermark? En sektor-ekonomisk analys" (What is the cost of maintaining arable land? A sectoral economic analysis). Unpublished manuscript. Uppsala: SLU.

Johnsen, Fred. 1990. "Økonomiske analyser av tiltak mot fosforavrenning fra dyrket mark" (Economic analysis of measures against phosphorus runoff from nonpoint agricultural sources). *Norwegian Agricultural Research,* Supplement no. 7:1990, pp. 9–118.

Johnson, D. Gale. 1984. "World Food and Agriculture." In Simon and Kahn, eds.: 67–112.

Johnsson, Bertil. 1990. "Mindre kväveläckage: Företagsekonomiska konsekvenser" (Less nitrogen leaching: Economic consequences). In Frank Petrini, ed., *Mot ett uthålligt och miljövänligt jordbruk.* Uppsala: Department of Economics, SLU.

Jonasson, Lars. 1989. *Intern avreglering och arealersättning* (Domestic deregulation and acreage payment). Report no. 21. Uppsala: Department of Economics, SLU.

——. 1991a. *Bonde på 90-talet* (Farmer in the 1990s). Aktuellt från lantbruksuniversitetet no. 391. Uppsala: SLU.

——. 1991b. *Effekter av miljöekonomiska styrmedel* (Effects of environmental policy instruments). Ekonomidagen 1991. Public report no. 175. Uppsala: SLU.

——. 1992. *Struktureffekter av skärpta miljökrav* (Structural effects of more stringent environmental demands). Småskrift Series no. 57. Uppsala: Department of Economics, SLU.

Junz, H. B., and Clemans Boonekamp. 1991. "What Is at Stake in the Uruguay Round?" *Finance and Development,* 28(2): 11–15.

Jägerhorn, Inger. 1992a. "Handeln sprider gifterna" (Commerce spreads poisons). *Dagens Nyheter,* 7 March. Stockholm.

——. 1992b. "Bondestöd sparas in" (Farmer support is reduced). *Dagens Nyheter,* 24 September. Stockholm.

Karlsson, Lars-Ingmar. 1992. "Biobränsle ger energi" (Biofuels give energy). *Dagens Nyheter,* 1 February. Stockholm.

Knutsen, Oddbjörn. 1989. "The Priorities of Materialist and Postmaterialist Values in the Nordic Countries." *Scandinavian Political Studies,* 12(3): 221–243.

Koester, Ulrich. 1991. "The Experience with Liberalization Policies: The Case of the Agricultural Sector." *European Economic Review,* 35(2/3): 562–570.

Kramer, Carol. 1991. "Implications of the Hormone Controversy for International Food Safety Standards." *Resources,* 105:12–14.

Kriström, Bengt. 1990. *Valuing Environmental Benefits Using the Contingent Valuation Method.* Economic Studies no. 219. University of Umeå.

Kumm, Karl-Ivar. 1984. "Gödsling, miljö och ekonomi: Kalkylexempel för gårdar i södra Halland," (Fertilization, environment and economy: An empirical example for fields in southern Halland). *Fakta/ekonomi,* 6. Uppsala: SLU.

——. 1989. *Skog på Jordbruksmark* (Forest on agricultural land). JTI report no. 105. Uppsala: Institute of Agricultural Technology, SLU.

Kumm, Karl-Ivar, and Rune Andersson. 1991. *Skog på Jorbruksmark: Ekonomi och miljö* (Forest on agricultural land: Economy and environment). Info Aktuellt no. 394. Uppsala: SLU.

Källander, Inger. 1991. "Uppvaktning hos jordbruksministern" (Calling on the Minister of Agriculture). *Alternativodlaren* (May) 37–38.

Lacy, William, and Lawrence Busch. 1989. "The Changing Division of Labor between the University and Industry: The Case of Agricultural Biotechnology." In Molnar and Kinnucan, eds.: 21–50.

Lacy, William, Laura Lacy, and Lawrence Busch. 1988. "Agricultural Biotechnology Research: Practices, Consequences, and Recommendations." *Agriculture and Human Values,* 5(3): 3–14.

Land. 1992. "Miljön vinnare i EG politik" (Environment the winner in EC policy). No. 23 (June).

Lawrence, Geoffrey. *1989.* "The Rural Crisis Down Under: Australia's Declining Fortunes in the Global Farm Economy." In Goodman and Redclift, eds.: 234–274.

Lerner, Thomas. *1992.* "Svensk ost chanslös" (Swedish cheese without luck). *Dagens Nyheter,* 30 May. Stockholm.

LeSS (Lantbruksekonomiska samarbetsnämndens sektorsgrupp) (Agricultural economics cooperative board's sector group). 1976. *Jordbrukets totalkalkyl 1970/71–1975/76* (Overall calculations for agriculture). Stockholm.

Lidén, Carl-Johan, and R. Andersson. 1989. *Legislation Measures for the Solving of Environmental Problems Resulting from Agricultural Practices, Their Economic Consequences, and the Impact on Agrarian Structures and Farm Rationalization.* Rome: UN Food and Agricultural Organization. FAO/ECE/Agri/WP.3/R.114.

Lindqvist, Sven. 1991. "Hur *A* blir ännu mera *B* än *B*" (How *A* became even more *B* than *B*). *Dagens Nyheter,* 24 May. Stockholm.

Lockeretz, William, and Molly Anderson. 1992. *Revitalizing Agricultural Research.* Lincoln: University of Nebraska Press.

LRF (Swedish Farmers' Federation). 1990a. *Lantbruksåret 1989* (The farm year 1989). Kalmar: LRF Information Annonsbyrån.

——. 1990b. *Yttrande över rapport angånde "En ny livsmedelspolitik"* (Statement on a new farm policy). DNR 1170/89. Stockholm.

——. 1992. "Struktur och strategiförutsättningar för den lantbrukarägda livsmedelsindustrin i ett gränslöst Europa" (Structural and strategic conditions for the farmer-owned food-processing industry in a borderless Europe). *Klara Fakta,* 3. Stockholm.

LSR (Lantbruksekonomiska samarbetsnämndens rationaliseringsgrupp) (Agricultural economics cooperative board's rationalization group). 1989. *Deltidlantbrukets struktur och betydelse* (Part-time farming's structure and significance). Partial report no. 2. Stockholm.

Lundgren, Lars. 1989. *Miljöpolitik på längden och tvären* (Environmental policy, lengthwise and crosswise). Report no. 3635. Solna: National Environmental Protection Board.

Lundin, Gunnar. 1988. *Ammoniak från stallgödsel* (Ammonia from manure). Report no. 94. Uppsala: Swedish Institute of Agricultural Engineering.

McCormack, Gavin. 1990. "Capitalism Triumphant? The Evidence from 'Number One.' Japan." *Monthly Review,* 42 (1): 1–13.

MacEwan, Arthur, and William Tabb, eds. 1990. *Instability and Change in the World Economy.* New York: Monthly Review Press.

McMichael, Philip, and David Myhre. 1990. "Global Regulation vs. the Nation-State: Agro-food Systems and the New Politics of Capital." *Review of Radical Political Economy,* 22(1): 59–77.

MacRae, Rod. 1989. *Policies, Programs, and Regulations to Support the Transition from Conventional to Sustainable Agriculture.* Report to Agriculture Canada. Montreal: Macdonald College of McGill University.

MacRae, Rod, Stuart Hill, John Henning, and Alison Bentley. 1990. "Policies, Programs, and Regulations to Support the Transition to Sustainable Agriculture in Canada." *American Journal of Alternative Agriculture,* 5(2): 76–92.

Malmström, Stefan. 1991. "Lagliga svin: Djurskydd och konkurrensförmåga" (Legal pigs: Animal protection and competitiveness). Degree essay no. 66. Uppsala: Department of Economics, SLU.

MCSA (Ministry for Civil Service Affairs). 1990. *Konkurrensen inom livsmedelssektorn* (Competition in the food sector). SOU 1990:25. Stockholm.

Micheletti, Michele. 1987. "Organization and Representation of Farmers' Interests in Sweden." *Sociologia Ruralis* 27(2/3): 166–180.

——. 1989. "The Swedish Elections of 1988." *Electoral Studies,* 8(2): 169–174.

——. 1990. *The Swedish Farmers' Movement and Government Agricultural Policy.* New York: Praeger.

Michélsen, Thomas. 1987a. "Gräsvallar bot mot kväveläckaget" (Grasslands work against nitrogen leaching). *Dagens Nyheter,* 6 June. Stockholm.

——. 1987b. "Gör sjuka djur olönsamma" (Make sick animals unprofitable). *Dagens Nyheter,* 10 June. Stockholm.

MOA (Ministry of Agriculture). 1946. *Riktlinjer för den framtida jordbrukspolitiken* (Directions for the future agricultural policy). SOU 1946:42. Stockholm.

——. 1977. Översyn av livsmedelspolitiken, mm (Overview of food policy). SOU 1977:17. Stockholm.

——. 1984. *Jordbruks- och livsmedelspolitik: Huvudbetänkande av 1983 års livsmedelskommitté* (Agricultural and food policy: Major considerations of the 1983 food commission). SOU 1984:86. Stockholm.

——. 1986. *Åtgärder för att minska spannmålsöverskottet* (Measures to reduce the excess production of grains). Ds Jo 1986:6. Stockholm.

——. 1987a. *Livsmedelspriser och livsmedelskvalitet.* SOU 1987:44. Stockholm.

——. 1987b. *Åtgärder för att minska spannålsöverskottet och stimulera alternativ markanvändning* (Measures to reduce excess grain production and stimulate alternative land use). Ds Jo 1987:2. Stockholm.

——. 1987c. *Intensiteten i jordbruksproduktionen* (Intensity in agricultural production). Ds Jo 1987:3. Stockholm.

——. 1989. *En ny livsmedelspolitik* (A new food policy). Ds Jo 1989:63. Stockholm.

——. 1990. *Pressmeddelande* (Press communication). No. 16-20, 1990-04-23. Stockholm.

——. 1991a. *Pressmeddelande* (Press communication). 1991-11-29. Stockholm.

——. 1991b. *Jordbruket i norra Sverige* (Agriculture in northern Sweden). SOU 1992:14. Stockholm.

——. 1992. "Mindre kadmium i handelsgödsel" (Less cadmium in fertilizer). SOU 1992:14. Stockholm.

——. 1993. *Jordbruksekonomiska meddelande* (Agricultural economics communication). No. 5. Stockholm.

MOE (Ministry of Environment and Energy). 1988. *Miljöpolitiken inför 90-talet* (Environmental policy on the eve of the 1990s). Stockholm.

——. 1990a. *Sätt värde på miljön! Miljöavgifter och andra ekonomiska styrmedel* (Put a value on the environment! Environmental taxes and other economic policy instruments). Final report of the Commission on Economic Instruments in Environmental Policy. SOU 1990:59. Stockholm.

——. 1990b. *Energipolitiken* (Energy policy). Government bill 1990/91:88. Stockholm.

——. 1991. *En god livsmiljö* (A good living environment). Government bill 1990/91:90. Stockholm.

MOF (Ministry of Finance). 1988. *Alternativ i jordbrukspolitiken* (Alternatives in agricultural policy). Ds 1988:54. Stockholm.

Molander, Per. 1988. *Säkerhetspolitiska aspekter på livsmedelsförsörjningen* (Security policy aspects of food provision). FOA report no. 10311-1.2. Stockholm: Swedish Defense Research Establishment.

——. 1992. "Jordbruksreformen myglas bort" (The agricultural reform is frittered away). *Dagens Nyheter,* 21 May. Stockholm.

Molnar, Joseph, and Henry Kinnucan, eds. 1989. *Biotechnology and the New Agricultural Revolution.* Boulder, Colo.: Westview Press.

Myrdal, Gunnar. 1938. *Jordbrukspolitiken under omläggning* (Agricultural policy during reorganization). Stockholm: Kooperativa Förbundets Bokförlag.

——. 1978. "Institutional Economics." *Journal of Economic Issues,* 12:771–783.

NAMB (National Agricultural Marketing Board). 1988a. *Vart går jordbruksstödet?* (Where does the agricultural support go?). Jönköping: Research Bureau.

——. 1988b. "Statistiskt meddelande" (Statistical message). Unpublished manuscript. Jönköping.

——. 1990. "Jordbruksekonomiska meddelanden" (Agricultural economics communication). Unpublished manuscript. Jönköping.

National Energy Board. 1989. *Ett miljöanpassat energisystem* (An environmentally adapted energy system). Report no. 3. Stockholm.

National Research Council. 1989. *Alternative Agriculture*. Washington, D.C.: National Academy of Sciences Press.

NBA (National Board of Agriculture). 1984. *Jordbrukets hänsyn till naturvårdens intressen* (Agriculture's responsibility for nature conservation interests). Public advisory no. 1984:1. Jönköping.

——. 1989a. *Minskad kemisk bekämpning* (Reduced chemical biocides). Jönköping.

——. 1989b. *Föreskrifter och allmänna råd om djurhållning inom lantbruket mm* (Instructions for animal husbandry in agriculture). LSFS 1989:20. Jönköping.

——. 1989c. *Nu gäller det Omställning 1990* (Now it concerns the 1990 adjustment). Lantbruksinformation 1989:15. Jönköping.

——. 1990a. *Lagring och spridning av stallgödsel* (Storage and spreading of manure). Report no. 1990:2. Jönköping.

——. 1990b. *Grön mark* (Green land). Report no. 1990:3. Jönköping.

——. 1990c. *Metoder att beräkna ekonomiska kosekvenser av djurskydds- och miljöbestämmelser* (Methods to calculate economic consequences of animal protection and environmental regulations). Report no. 1990:1. Jönköping.

——. 1990d. *Åtgärder enligt handlingsprogrammet för att minska hälso- och miljöriskerna vid användning av bekämpningsmedel* (Measures according to the action program to reduce health and environmental risks from use of biocides). Report no. 1990:8. Jönköping.

——. 1990e. *Skötsellagen* (The Land Care Law). NBA's public advisory no. 1990:1. Jönköping.

——. 1991a. *Sammanställning av effekterna av pågående och förslaget åtgärdsprogram för att minska jordbrukets växtnäringsförluster* (Compilation of the effects of current and recommended measures to reduce agricultural nutrient losses). Document no. 34210/91. Jönköping.

——. 1991b. *Resultat av handlingsprogrammet för att minska hälso- och miljöriskerna vid användning av bekämpningsmedel* (Results of the program to reduce health and environmental risks of using biocides). Report no. 1991:8. Jönköping.

——. 1991c. *Svenskt jordbruk i EG* (Swedish agriculture in the EC). Report no. 1991:5. Jönköping.

NCFAP (National Center for Food and Agricultural Policy). 1991. *Should Agriculture Go with the GATT?* Washington, D.C.: Briefing Book.

Nello, Susan. 1989. "European Interest Groups and the CAP." *Food Policy*, (2): 101–106.

NEPB (National Environmental Protection Board). 1986. *Jordbruket och miljön* (Agriculture and the environment). Solna.

——. 1989. *Livsmedelstoxikologiska aspekter på rötslam i jordbruket* (Food toxological aspects of sludge in agriculture). Report no. 3623. Solna.

——. 1990a. *Hav '90: Aktionsprogram mot havsföroreningar* (Ocean '90: Action Program against Ocean Pollutants). Solna.

——. 1990b. *Slam från kommunala avloppsreningsverk* (Sewage from municipal treatment plants). Public advisory no. 1990:13. Solna.

——. 1990c. *Natur 90: Aktionsprogram för naturvård* (1990 program for nature conservation). Solna.

——. 1990d. *An Environmentally Adapted Energy System.* Report no. 3744. Stockholm.
——. 1991. *Länsvisa program för bevarande av odlingslandskapets natur och kulturvärden* (Regional programs to preserve the nature and cultural values of the agricultural landscape). Public advisory no. 91:3. Solna.
Nielsen, Klaus, and Ove Pedersen. 1990. "From the Mixed Economy to the Negotiated Economy." Paper presented at the conference of the Society for Advancement of Socio-economics, March, Washington, D.C.
Nilson, Astrid. 1992. "Billigare mat efter 12 års prishöjning" (Cheaper food after 12 years of price increases). *Dagens Nyheter,* 3 October. Stockholm.
Nilsson, Ingemar. 1989. "The Effects on Swedish Agriculture of Conversion to EC and GATT Regulations as well as New Environmental Demands" (in Swedish). In E.F. og Miljø Jordbrug. Copenhagen: Royal Seminar, pp. 1–7.
Nilsson, Magnus. 1988. *Fingret på voteringsknappen* (Finger on the voting button). Stockholm: Swedish Society for Conservation of Nature.
Nitsch, Ulrich. 1991. "Knowledge of Agriculture." Paper presented at the conference of the Sociological Society, Columbus, Ohio.
NM (Nordic Ministerial Council). 1989. *Jordbrukspolitiken i de nordiska länderna* (Agricultural policy in the Nordic countries). NORD report no. 1989:15. Copenhagen: Nordic Ministerial Council.
Nordhaus, William. 1991. "To Slow or Not to Slow: The Economics of the Greenhouse Effect." *Economic Journal,* 101:920–937.
Norgaard, Richard. 1991. "Sustainability: The Paradigmatic Challenge to Agricultural Economics." Paper presented at the 21st Conference of the International Association of Agricultural Economists, Tokyo.
Nyberg, Jan Erik. 1989. "Ett efterfrågestyrt jordbruk" (A demand-steered agriculture). In Tord Eriksson, ed., *Framtidens jordbrukspolitik* (The future's agricultural policy). Stockholm: Royal Forestry and Agricultural Academy, pp. 44–47.
OECD (Organization for Economic Cooperation and Development). 1987a. *National Policies and Agricultural Trade.* Paris: OECD.
——. 1987b. "OECD Farmers and Agricultural Policies: The Costs of Oversupply." *OECD Observer,* 147:5–9. Paris.
——. 1987c. *National Policies and Agricultural Trade. Country Study: United States.* Paris.
——. 1988a. *National Policies and Agricultural Trade. Country Study: Sweden.* Paris.
——. 1988b. "Agricultural Reform: A Long Row to Hoe." *OECD Observer* (June–July): 16–19. Paris.
——. 1989a. "OECD in Figures: Statistics of the Member Countries." *OECD Observer, Supplement* 158. Paris.
——. 1989b. *Agriculture and Environmental Policies.* Paris.
——. 1991a. "Agricultural Policy and the Urgency of Reform." *OECD Observer,* 168:30–35. Paris.
——. 1991b. *National Policies and Agricultural Trade: Country Study: Sweden.* Paris.
——. 1992. *Agricultural Policies, Markets, and Trade.* Paris.
Olingsberg, Thomas. 1990. "Förnedrande väntan på kronofogden" (Debasing wait for the debt collector). *Dagens Nyheter,* 28 July. Stockholm.

Olson, Mancur. 1965. *The Logic of Collective Action.* Cambridge, Mass.: Harvard University Press.

——. 1982. *The Rise and Decline of Nations.* New Haven: Yale University Press.

Olsson, Håkan, and Stefan Löfgren. 1989. *Tillförsel av kväve och fosfor till havet* (Discharge of nitrogen and phosphorus to the sea). Report no. 3693. Stockholm: National Environmental Protection Board.

Østergaard, V. 1989. "Establishment of Organic Experimental Farms in Denmark." In Dubgaard and Nielsen, eds.: 289–298.

Ostry, Sylvia. 1991. "The Uruguay Round: An Unfinished Symphony." *Finance and Development,* 28:2.

Parikka, Matti. 1989. *Energiskogens ekonomi* (The economics of fast-growing trees for energy production). Essay series no. 31. Uppsala: Department of Forest Industry and Market Studies, SLU.

Passell, Peter. 1992. "Tuna and Trade: Who's Rules?" *New York Times,* 19 February.

Pestoff, Victor. 1990. "Organizing Consumers in a Negotiated Economy: Swedish Consumer Policy." Paper presented at the conference of the Society for Advancement of Socio-economics, March, Washington, D.C.

——. 1991. "The Demise of the Swedish Model and the Rise of Organized Business as a Major Political Actor." Paper presented at the conference of the Advancement of Socio-economics, June, Stockholm.

Peterson, Wesley, and Clare Lyons. 1989. "The Perpetual Agricultural Crisis in the European Community." *Agriculture and Human Values,* 7 (1 & 2): 11–21.

Petersson, Olof, and Ingrid Carlberg. 1990. *Makten över tanken* (Power over thought). Stockholm: Carlssons Bokförlag.

Petit, Michel. 1985. *Determinants of agricultural policies in the United States and the European Community.* Research report no. 51. Washington, D.C.: International Food Policy Research Institute.

Pettersson, Conny. 1990. "Inget bete för kossor på Hamra: Alfa-Laval får dispens" (No grazing for cows in Hamra: Alfa-Laval gets an exemption). *Dagens Nyheter,* 30 June. Stockholm.

Pettersson, Olle. 1989. "Hur kom miljöforskningen till lantbrukshögskolan?" (How did environmental research come to the agricultural university?). *Skogs och Lantbruksakadamiens Tidskrift,* 128:191–207.

Pevetz, I. 1990. "Agriculture and Environmental Protection in Austria." Wageningen: Working Party on Agrarian Structure and Farm Rationalization. UN FAO Economic Commission for Europe. FAO/ECE/Agri/WP.3/R.117.

Pollin, Robert. 1990. "Debt-dependent Growth and Financial Innovation." In MacEwan and Tabb, eds.: 121–146.

Porter, Michael. 1990. *The Competitive Advantage of Nations.* New York: Free Press.

Rabe, Gunnar. 1990. *Skattereformen* (The tax reform). Stockholm: Föreningen för Auktoriserade Revisioner.

Rabinowicz, Ewa. 1991a. *Market Power and Consumer Welfare: Agricultural Policy—Old Wine in New Bottles.* Study of Power and Democracy in Sweden, Report no. 32. Stockholm: SNS Förlag.

——. 1991b. "Ny omställning väntar svenska lantbruket" (New adjustments await Swedish agriculture). *Svenska Dagbladet,* 13 August. Stockholm.

Rabinowicz, Ewa, Olof Bolin, and Ingemar Haraldsson. 1986. "The Evolution of a Regulation System in Agriculture: The Swedish Case." *Food Policy,* 11(4): 323–333.

Randall, Allen. 1987. *Resource Economics.* 2d ed. New York: Wiley.

Rauch, Jonathan. 1991. "Spilled Milk." *National Journal,* 14 September, pp. 2210–2214.

Reichelderfer, Katherine. 1990. "Environmental Protection and Agricultural Support: Are Tradeoffs Necessary?" In Kristen Allen, ed., *Agricultural Policies in a New Decade.* Washington, D.C.: Resources for the Future, pp. 201–229.

——. 1991. "The Expanding Role of Environmental Interests in Agricultural Policy." *Resources,* 102:4–7. Washington, D.C.: Resources for the Future.

Revelle, Roger. 1984. "The World Supply of Agricultural Land." In Simon and Kahn, eds.: 182–201.

Riemenschneider, Charles, and Robert Young II. 1989. "Agriculture and the Failure of the Budget Process." In Carol Kramer, ed., *The Political Economy of U.S. Agriculture: Challenges for the 1990s.* Washington, D.C.: Resources for the Future, pp. 87–104.

Rothstein, Bo. 1988. *Social Classes and Political Institutions: The Roots of Swedish Corporatism.* Study of Power and Democracy in Sweden, Report no. 24. Uppsala.

——. 1991. "De borgerliga får skylla sig själva" (The bourgeoisie can blame itself). *Dagens Nyheter,* 10 June. Stockholm.

——. 1992. "The Crisis of the Swedish Social Democrats and the Future of the Universal Welfare State." Unpublished manuscript. Cambridge, Mass.: Center for European Studies, Harvard University.

Rowinski, Tadeusz, and Bertil Johnsson. 1988. *Ekonomisk analys av några inhysningssystem för värphöns* (Economic analysis of a few housing systems for laying hens). Report no. 21. Uppsala: Department of Economics, SLU.

——. 1990. *Ekonomisk analys av några skötselsystem för slaktsvin* (Economic analysis of a few husbandry systems for swine). Report no. 26. Uppsala: Department of Economics, SLU.

Runge, C. Ford, and R. Nolan. 1990. "Trade in Disservices." *Food Policy,* 15(1): 3–7.

Ruttan, Vernon. 1990. "Scientific, Technical, Resource, Environmental, and Health Constraints on Sustainable Growth in Agricultural Production: Into the 21st Century." Paper presented at the Symposium on Population-Environment Dynamics, University of Michigan, Ann Arbor.

Sagoff, Mark. 1988. "Biotechnology and the Environment: What Is at Risk?" *Agriculture and Human Values,* 5(3): 26–35.

Sandberg, Peter. 1990. "Kväve från floderna bäddar för giftalger" (Nitrogen from the rivers, a bath for poison algae). *Dagens Nyheter,* 10 February. Stockholm.

SBA (Swedish Board of Agriculture). 1991a. *Utvärdering av den livsmedelspolitiska reformen* (Evaluation of the food policy reform). Report no. 1991:4. Jönköping.

——. 1991b. *Marknadsutveckling* (Market development). Report no. 1991:5. Jönköping.

——. 1991c. *Utvärdering av omställningsåtgärder mm i den nya livsmedelspoli-*

tiken (Evaluation of transition measures, etc., in the new food policy). Report no. 1991:6. Jönköping.

——. 1991d. *Strukturutvecklingen i jordbruket* (Structural changes in agriculture). Report no. 1991:7. Jönköping.

——. 1991e. *Ammoniakförluster i jordbruket: Förslag till åtgärdsprogram* (Ammonia emissions in agriculture: Proposal for an action program). Report no. 1991:11. Jönköping.

——. 1992a. *Omställningsarealernas användning* (The use of converted acreage). Report no. 1992:15. Jönköping.

——. 1992b. *Jordbruket i EG: EGs jordbruksreform* (Agriculture in the EC: The EC's agricultural reform). Report no. 1992:13. Jönköping.

——. 1992c. *Miljöavgifter: bekämpningsmedel och handelsgödsel* (Environmental taxes on chemical biocides and commercial fertilizer). Report no. 1992:41. Jönköping.

SCEES (Service Central des Enquêtes et Etudes Statistiques). 1991. *GRAFAGRI 91*. Paris: Agreste mai 91, Ministère de l'Agriculture et de la Forêt.

Schoemaker, Robin. 1989. "Agricultural Land Values and Rents under the Conservation Reserve Program." *Land Economics*, 65(2): 131–137.

Sedjo, Roger. 1989. "Forests to Offset the Greenhouse Effect." *Journal of Forestry*, 87(7): 12–15.

Sharples, Jerry, and Janette Krutzfeldt. 1990. *World Grain Stocks*. Bulletin 594. Economic Research Service, USDA. March.

SIAE (Swedish Institute of Agricultural Engineering). 1984. *Bränslen från jordbruket* (Fuel from agriculture). Stage Four report. Uppsala: Agrobioenergy Project.

SIFO (Swedish Institute for Opinion Research). 1989a. *Det naturliga steget: Eftermätning* (The natural step: Measurement). Stockholm.

——. 1989b. "SIFO om miljöopinionen" (SIFO on environmental opinion). *Miljö-Rapporten*. No. 2, Stockholm, December, p.7.

Silk, Leonard. 1992. "Failure Looming for Trade Talks." *New York Times*, 28 February.

Silvander, Ulf. 1991. *Betalningsvillighetsstudier för sportfiske och grundvatten* (Studies on willingness to pay for angling and groundwater). Uppsala: Department of Economics, SLU.

——. 1993. "Externa effekter av vallproduktion för biogasframställning" (External effects of grass production for biogas: Presentation). Småskrifts Series no. 70. Uppsala: Department of Economics, SLU.

Silvander, Ulf, and Lars Drake. 1990. *Ekonometrisk analys av efterfrågan på vissa basvaror* (Econometric analysis of demand for certain basic foods). Report no. 20. Uppsala: Department of Economics, SLU.

——. 1991. "Nitrate Pollution and Fisheries Protection in Sweden." In Hanley, ed.: 159–178.

Simon, Julian, and Herman Kahn, eds. 1984. *The Resourceful Earth*. Oxford: Blackwell.

Simons, Marlise. 1992. "Alps Caught in Vise between Tourism and Trucks." *New York Times*, 6 April.

Sjöström, Alf. 1989. *Valet 1988* (1988 election). Stockholm: SIFO.

SNF (Svenska Naturskyddsförening, Swedish Society for Nature Conservation).

1990. *Remissvar till "En ny livsmedelspolitik"* (Remiss response to "A new food policy"). SNF Dnr 315/89. Stockholm.

Sødal, D. 1989. "Regional Employment Effects of the New Measure to Control Pollution from Animal Husbandry in Norway." In Dubgaard and Nielsen, eds.: 177–186.

Sohlman, Michael. 1989. "Miljö- och jordbrukspolitik i Sverige" (Environmental and agricultural policy in Sweden). In *Jordbrug, EF og Miljø*. Copenhagen: Royal Seminar, pp. *1–9*.

Southgate, Douglas, J. Sanders, and S. Ehui. *1990*. "Resource Degradation in Africa and Latin America: Population Pressure, Policies, and Property Arrangements." *American Journal of Agricultural Economics,* 72(5): 1259–1263.

SP (Swedish Parliament). 1990. *Jordbruksutskottets betänkande: Livsmedelspolitiken* (Permanent Agricultural Committee's thinking: Food policy). JOU 1989/90:25. Stockholm.

SPA (Government Planning Administration). 1983. *Riksplaneyttrande 1983* (1983 pronouncement on national physical planning). Communication no. 39/83. Stockholm.

SS (Statistics Sweden). 1982. *Tätortsexpansion på jordbruksmark 1975–1980* (Urban expansion on arable land, 1975–80). SM J 1982:10. Stockholm.

——. 1990a. *Jordbruksstatistisk Årsbok 1990* (Yearbook of agricultural statistics 1990). Stockholm.

——. 1990b. "Naturmiljön i siffror" (The natural environment in figures). Stockholm.

——. 1991a. *Jordbruksstatistisk årsbok 1991* (Yearbook of agricultural statistics 1991). Stockholm.

——. 1991b. "Jordbruksekonomiska undersökningen 1990" (The 1990 farm economy survey). *Statistiska Meddelanden* J 40 SM 9101. Örebro.

——. 1991c. *Miljösverige* (Environment Sweden). Stockholm.

——. 1992. *Jordbruksstatistisk Årsbok 1992* (Yearbook of Agricultural Statistics 1992). Stockholm.

Stern, Paul. 1990. "The Social Construction of the Economy." *Challenge* (January–February): 38–45.

Sterner, Frank. 1990. "Fånggrödan i jordbruksföretaget: Kvävefångst utan ekonomikvävning?" (Catch crops in the agricultural enterprise: A nitrogen trap without economic strangulation?). *Fakta, Ekonomi* 3. Uppsala: Department of Economics, SLU.

Sundell, Björn. 1982. *"Ekonomiska studier av nytta och risk vid bekämpning av växtskadegörare"* (Economic studies of benefits and risks of controlling weed pests). Master's thesis. Department of Economics and Statistics, SLU, Uppsala.

Swedish Council for Forest and Agricultural Research. 1989. *Minskad bekämpning i jordbruket: Möjligheter och konsekvenser* (Reduced pesticide use in agriculture: Possibilities and consequences). Report no. 36. Stockholm.

Swedish Institute. 1988. *Agriculture in Sweden*. Stockholm.

——. 1991. "Energy and Energy Policy in Sweden." FS 37. Stockholm: Fact Sheets on Sweden.

Tamminga, Gustaaf, and Jo Wijnands. 1991. "Animal Waste Problems in the Netherlands." In Hanley, ed.: 117–136.

Taylor, Robert. 1991. "Sweden's Recovery Seen as Slow and Possibly Short-lived." *Financial Times,* 16 July.

——. 1992a. "Swedish Industry Says Enough Is Enough" *Financial Times,* 5 February, p. 19.

——. 1992b. "Swedes Boost Defence Sector." *Financial Times,* 26 February, p. 2.

Thelander, Jan, and Lars Lundgren. 1989. *Nedräkning Pågår* (The countdown continues). Solna: NEPB.

Thullberg, Per. 1983. "Swedish Agricultural Policy during the Postwar Period: A Study in Rationality." Unpublished manuscript. Florence: European Consortium for Political Research.

Tobey, James. 1991. "The Effects of Environmental Policy towards Agriculture on Trade." *Food Policy,* 16(2): 90–94.

Tobey, James, and David Ervin. 1990. "Environmental Considerations of the Single European Act." Paper presented at the conference "Europe 1992: The Future for World Agriculture." Washington, D.C.: USDA.

Uhlin, Hans-Erik. 1989. *Jordbrukspolitiska frågor inför 2000-talet* (Agricultural policy questions on the eve of the 21st century). Report no. 37. Stockholm: Royal Forestry and Agricultural Academy.

Uimonen, Peter. 1992. "Trade Policies and the Environment." *Finance and Development,* 29(2): 26–27.

UIPP (Union des Industries de la Protection des Plantes). 1991. Manuscript of unpublished statistics. Paris.

USDA (U.S. Department of Agriculture). 1989. *Agricultural Statistics 1988.* Washington, D.C.

——. 1990a. *Agricultural Resources—Inputs: Situation and Outlook Report.* AR-21. Economic Research Service.

——. 1990b. *Agricultural Outlook.* AO-161. March.

——. 1990c. *Milk Production.* February.

——. 1990d. "U.S. Farm Sector Weathers Storm, Builds Strength." *Farmline* (January): 4–7.

——. 1991. *Agricultural Outlook.* AO-176. June.

Vail, David. 1993. "The Devolution of the Negotiated Economy? Sweden's 1990 Food Policy Reform." In Sven-Erik Sjöstrand, ed., *Institutional Change: Theory and Empirical Findings.* Armonk, N.Y.: Sharpe, pp. 301–317.

——. Forthcoming. "From Democratic Corporatism to Neo-liberalism: The Domestic and European Context of Sweden's 1990 Food Policy Reform." In Philip McMichael, ed., *Restructuring Agricultural and Food Systems in the Late Twentieth Century.* Ithaca, N.Y.: Cornell University Press.

van den Berg, Leo. 1988. "Rural Land-use Planning in the Netherlands: Integration or Segregation of Functions?" In Cloke, ed.: 47–74.

van den Weghe, H. 1990. "Animal Waste and the Environment." Wageningen: Economic Commission for Europe, Working Party of Agrarian Structure and Farm Rationalization, UN FAO. FAO/ECE/AGRI/WP.3/R.110.

van Leeuwen, G., and H. Oosterveld. 1989. "Agriculture and Environment in the Netherlands: Problems, Measures, and Solutions." Wageningen: Economic Commission for Europe, UN FAO. FAO/ECE/AGRI/WP.3/R.116.

van Lier, Hubert. 1988. "Land-use Planning on Its Way to Environmental Planning." In Whitby and Ollerenshaw, eds.: 89–117.

Vatn, Arild. 1989. "Agricultural Policy, Regional Specialization, and Environmental Problems in Norway." In Dubgaard and Nielsen, eds.: 167–176.

Viatte, Gérard, and Carmel Cahill, 1991. "The Resistance to Agricultural Reform." *OECD Observer,* 171 (August–September), Paris, pp. 4–8.

Viatte, Gérard, and Frédéric Langer. 1990. "Agricultural Reform: A Hesitant Start." *OECD Observer,* 165:4–8. Paris.

WCED (World Commission on Environment and Development). 1987. *Our Common Future.* New York: Oxford University Press.

Weaver, R. Kent. 1987. "Political Foundations of Swedish Economic Policy." In B. Bosworth and A. Rivlin, eds., *The Swedish Economy.* Washington, D.C.: Brookings, pp. 289–324.

Weinschenck, G., and S. Dabbert. 1989. "Decrease in the Intensity of Natural Resource Use as a Way to Reduce Surplus Production." In Dubgaard and Neilsen, eds.: 25–34.

Westberg, Lotten. 1988. "Lantbruket i dagspressen: En innehållsanalys" (Farming in the daily press: A content analysis). Master's thesis. Department of Extension Education, SLU, Uppsala.

Whitby, Martin, and John Ollerenshaw, eds. 1988. *Land Use and the European Environment.* London: Belhaven Press.

Winter, L. Alan. 1987. "The Political Economy of Agricultural Policy of Industrial Countries." *European Review of Agricultural Economics,* 14:285–304.

World Bank. 1986. *World Development Report 1986.* New York: Oxford University Press.

Young, Douglas. 1989. "Policy Barriers to Sustainable Agriculture." *American Journal of Alternative Agriculture,* 4(4): 135–143.

Young, M. D. 1988. "Some Steps in Other Countries." *EPA Journal,* 14(3): 24–25.

——. 1989. "The Integration of Agricultural and Environmental Policies." In Dubgaard and Nielsen, eds.: 3–14.

Åmark, Karl. 1942. "Svensk jordbrukspolitik under det senaste kvartsseklet" (Swedish agricultural policy during the past quarter century). In *Sveriges lantbruksförbund 1917–42.* Stockholm: Lantbruksförbundets Tidskriftsaktiebolag.

Index

Food Systems and Agrarian Change

Edited by Frederick H. Buttel, Billie R. DeWalt, and Per Pinstrup-Andersen

Hungry Dreams: The Failure of Food Policy in Revolutionary Nicaragua, 1979–1990
by Brizio N. Biondi-Morra

Research and Productivity in Asian Agriculture
by Robert E. Evenson and Carl E. Pray

The Politics of Food in Mexico: State Power and Social Mobilization
by Jonathan Fox

Searching for Rural Development: Labor Migration and Employment in Mexico
by Merilee S. Grindle

The Global Restructuring of Agro-Food Systems
edited by Philip McMichael

Structural Change and Small-Farm Agriculture in Northwest Portugal
by Eric Monke et al.

Diversity, Farmer Knowledge, and Sustainability
edited by Joyce Lewinger Moock and Robert E. Rhoades

Networking in International Agricultural Research
by Donald L. Plucknett, Nigel J. H. Smith, and Selcuk Ozgediz

Adjusting to Policy Failure in African Economies
edited by David E. Sahn

The New Economics of India's Green Revolution: Income and Employment Diffusion in Uttar Pradesh
by Rita Sharma and Thomas T. Poleman

Agriculture and the State: Growth, Employment, and Poverty in Developing Countries
edited by C. Peter Timmer

The Greening of Agricultural Policy in Industrial Societies: Swedish Reforms in Comparative Perspective
by David Vail, Knut Per Hasund, and Lars Drake

Transforming Agriculture in Taiwan: The Experience of the Joint Commission on Rural Reconstruction
by Joseph A. Yager